ISSN 0044-2267

ZEITSCHRIFT FÜR ANGEWANDTE MATHEMATIK UND MECHANIK

INGENIEURWISSENSCHAFTLICHE FORSCHUNGSARBEITEN

BAND 62 1982 HEFT 6

AUS DEM INHALT

HAUPTAUFSÄTZE

AKADEMIE-VERLAG · BERLIN

ZAMM Bd. 62 Nr. 6 S. 213—284 Berlin, Juni 1982

EVP 18,— M 34115

INHALT

Hauptaufsätze

Wir bitten, **Manuskriptsendungen** zweifach (Original und eine Kopie, sprachlich einwandfrei, Formeln mit Maschine oder in Druckschrift geschrieben) an folgende Anschrift zu richten:

Zeitschrift für Angewandte Mathematik und Mechanik, z. Hd. Herrn Prof. Dr. Günter Schmidt
Institut für Mechanik der Akademie der Wissenschaften der DDR
DDR-1199 Berlin, Rudower Chaussee 5

Zu den Arbeiten, die als Hauptaufsätze bestimmt sind, ist auf gesondertem Blatt eine Zusammenfassung von 5 bis 10 Zeilen in englischer und (möglichst) deutscher und russischer Sprache beizufügen. Ausführliche Hinweise für die Autoren, um deren strikte Berücksichtigung gebeten wird, finden sich im Anschluß an das Inhaltsverzeichnis des letzten Jahrganges.

Die Autoren erhalten von den Hauptaufsätzen 75, von den Kleinen Mitteilungen 25 Sonderdrucke ohne Berechnung, darüber hinaus weitere Sonderdrucke gegen Berechnung.

Der Verlag behält sich für alle Beiträge das Recht der Vervielfältigung und Übersetzung vor.

Bestellungen sind zu richten

- **in der DDR** an den Postzeitungsvertrieb unter Angabe der Kundennummer des Bestellenden oder an den AKADEMIE-VERLAG, DDR-1086 Berlin, Leipziger Str. 3—4, PSF 1233
- **im sozialistischen Ausland** an eine Buchhandlung für fremdsprachige Literatur oder an den zuständigen Postzeitungsvertrieb
- **in der BRD und Berlin(West)** an eine Buchhandlung oder an die Auslieferungsstelle KUNST UND WISSEN, Erich Bieber OHG, D-7000 Stuttgart 1, Wilhelmstraße 4—6
- **in den übrigen westeuropäischen Ländern** an eine Buchhandlung oder an die Auslieferungsstelle KUNST UND WISSEN, Erich Bieber GmbH, CH-8008 Zürich, Dufourstraße 51
- **im übrigen Ausland** an den Internationalen Buch- und Zeitschriftenhandel; den Buchexport, Volkseigener Außenhandelsbetrieb der Deutschen Demokratischen Republik, DDR-7010 Leipzig, Postfach 160; oder an den AKADEMIE-VERLAG, DDR-1086 Berlin, Leipziger Straße 3—4, PSF 1233

ZEITSCHRIFT FÜR ANGEWANDTE MATHEMATIK UND MECHANIK

Herausgeber und Chefredakteur: Prof. Dr. Günter Schmidt. Redaktion: Dr. Winfried Heinrich, Dr. Horst Weinert, Dipl.-Math. Friedhild Dudel, Bernhilde Schwertfeger, Institut für Mechanik der Akademie der Wissenschaften der DDR. Verlag: Akademie-Verlag, DDR-1086 Berlin, Leipziger Straße 3—4; Fernruf: 2236221 oder 2236229; Telex-Nr.: 114420; Bank: Staatsbank der DDR, Berlin, Kto.-Nr. 6836-26-20712. Anschrift der Redaktion: Institut für Mechanik der Akademie der Wissenschaften, DDR-1199 Berlin, Rudower Chaussee 5; Fernruf: 6743639 oder 6743643.

Veröffentlicht unter der Lizenznummer 1282 des Presseamtes beim Vorsitzenden des Ministerrates der Deutschen Demokratischen Republik. Gesamtherstellung: VEB Druckerei „Thomas Müntzer", 5820 Bad Langensalza. Erscheinungsweise: Die Zeitschrift für Angewandte Mathematik und Mechanik erscheint monatlich. Die 12 Hefte eines Jahres einschließlich Tagungsheft bilden einen Band. Bezugspreis je Band 360,— M zuzüglich Versandspesen (Preis für die DDR 216,— M). Bezugspreis je Heft 30,— M (Preis für die DDR 18,— M). Bestellnummer dieses Heftes: 1009/62/6.

AN (EDV) 35937

ZAMM 62, 213 – 219 (1982)

J. N. Reddy / C. W. Bert

On the Behavior of Plates Laminated of Bimodulus Composite Materials

An exact and finite-element analysis is used to predict the static bending and natural vibration response of rectangular plates laminated of composite material having different elastic properties depending upon whether the fiber-direction strains are tensile or compressive. The analysis is based on a theory that accounts for anisotropy, bending-stretching coupling, thickness shear deformation, and both coupling inertia and rotatory inertia. The finite-element results are found to agree very closely with the exact closed-form solutions in the case of a cross-ply rectangular plate having simply supported edges and subjected to sinusoidally distributed normal-pressure loading or temperature loading.

Um statische Biegung und Eigenschwingungsantwort von geschichteten Rechteckplatten vorauszusagen, die aus zusammengesetztem Material bestehen, das unterschiedliche elastische Eigenschaften besitzt, je nachdem, ob die Dehnung in Faserrichtung Druck- oder Zugcharakter hat, wird eine exakte wie auch eine Finite-Elemente-Analyse angewandt. Die Rechnung basiert auf einer Theorie, die Anisotropie, Biegungs-Dehnungs-Kopplung, Schubdeformation über die Dicke und sowohl Kopplungs- als auch Verdrehungsträgheit berücksichtigt. Es ergibt sich, daß die Ergebnisse aus der Finite-Elemente-Analyse im Falle einer kreuzweise beschichteten Rechteckplatte mit einfach unterstützten Kanten, die sinusoidal-verteilten Normaldruck- oder Temperaturbelastungen ausgesetzt ist, sehr gut mit den exakten Lösungen in geschlossener Form übereinstimmen.

Точный анализ и анализ с помощью метода финитных элементов используются для определения статического изгиба и собственных колебаний слоистых прямоугольных пластин, состоящих из композиционного материала, имеющего различные упругие свойства в зависимости от того, является ли напряжение по направлению волокон растягивающим или сжимающим. Анализ основан на теории, учитывающей анизотропию, связь изгиб-растяжение, деформации сдвига по толщине и инерции связи и вращения. Результаты, найденные с помощью метода финитных элементов, очень близки к точным решениям в замкнутой форме в случае специальной прямоугольной пластины, просто опёртой на краях и испытывающей синусоидально распределённую нагрузку нормального давления или температуры.

1. Introduction

Certain fiber-reinforced materials, especially those with very soft matrix material (e.g., cord-rubber composites), exhibit different elastic behavior depending upon whether the fiber-direction strain is tensile or compressive [1—3]. In other words, the tangent modulus in tension is quite different from the tangent modulus in compression. As a first approximation the uniaxial stress-strain behavior of such materials is often represented as being linear with different slopes (i.e., elastic moduli) depending upon the sign of the fiber-direction strain. Such a material is called a *bimodulus* composite material. Several macroscopic material models appropriate for bimodulus fiber-reinforced composites have been proposed and are reviewed in [4]. It has been shown that the fiber-governed symmetric-compliance model proposed in [5] agrees well with experimental data for several materials with drastically different elastic properties in tension and compression.

Prior to the works of the present authors, the literature available in English on bending and vibration analyses of bimodulus plates was quite sparse and largely concerned with the bending of bimodulus isotropic-material plates. Shapiro [6] considered the simple problem of a circular plate under a pure bending moment at its edge. Kamiya [7] analyzed the large-deflection behavior of clamped circular plates using a finite-difference technique. In [8], Kamiya used the Galerkin method to analyze the large-deflection behavior of rectangular plates under sinusoidally distributed loading. The effect of thickness shear deformation was included in the simple case of cylindrical bending by Kamiya [9]. The only previous analyses applicable to anisotropic bimodulus material are the works of Jones and Morgan [10], who treated cylindrial bending of a thin, cross-ply laminate, and the closed-form solutions of Kincannon and Bert [11, 12] for clamped elliptic plates. All of the previous thermoelastic analyses of bimodulus-material plates were also limited to isotropic bimodulus materials and midplane-symmetric temperature changes [13—17]. Apparently, there were no published papers on the vibration of bimodulus plates.

In the last two years, the present authors have contributed a series of papers to the literature on bimodulus rectangular plates [18—23]. These works are apparently more general than any analyses that have appeared in the open literature. More specifically, these works differ from previous works in the following respects:

1. The material of each layer is both elastically and thermoelastically orthotropic and bimodular.
2. Both single-layer orthotropic and two-layer cross-ply laminated plate constructions are considered.
3. Thickness shear deformations are included.
4. Temperature changes through the thickness as well as in the plane are considered.
5. Both finite-element and exact closed-form solutions are presented.

The objective of the present paper is two-fold: to review the recent developments in the analyses of bimodulus-material plates with finite dimensions and to present new finite-element results for rectangular plates under conditions that do not admit closed-form solutions.

2. Governing Equations for Bimodulus Plates

In the following the equations governing the shear deformable theory of layered composite plates (see Whitney and Pagano [24]) is reviewed and the thermoelastic constitutive equations of bimodulus-material plates are presented.

In deriving the governing equations it is assumed that all of the layers in the plate remain elastic during the deformation, the generalized HOOKE's law is valid, and no slip occurs between any two layers.

Consider a plate constructed of a finite number of uniform thickness, orthotropic, bimodulus layers with arbitrary orientations (i.e., the material symmetry axes of each layer do not coincide, in general with the plate axes). The x-y plane lies in the middle plane, R, of the plate and the z-axis is normal to R. The displacement field in the shear deformable theory of plates is assumed to be given by

$$u(x, y, z, t) = u_0(x, y, t) + z\psi_x(x, y, t)\,, \quad v(x, y, z, t) = v_0(x, y, t) + z\psi_y(x, y, t)\,, \quad w = w(x, y, t) \tag{2.1}$$

where u, v, and w are the displacements along x, y and z directions, respectively, u_0 and v_0 are the inplane displacements of the middle plane, ψ_x and ψ_y are the bending slopes, and t is time. Using the small-deflection theory, the strain-displacement relations can be expressed in the form

$$\begin{aligned} &\varepsilon_1 \equiv \varepsilon_x = u_{0,x} + z\psi_{x,x}\,, \quad \varepsilon_2 \equiv \varepsilon_y = v_{0,y} + z\psi_{y,y}\,, \quad \varepsilon_6 \equiv \gamma_{xy} = u_{0,y} + v_{0,x} + z(\psi_{x,y} + \psi_{y,x})\,, \\ &\gamma_{xz} = \psi_x + w_{,x}\,, \quad \gamma_{yz} = \psi_y + w_{,y}\,, \quad \varepsilon_3 = 0\,, \end{aligned} \tag{2.2}$$

where $w_{,x} \equiv \partial w/\partial x$, etc.

The equations of motion are given by

$$\begin{aligned} &N_{1,x} + N_{6,y} = Pu_{0,tt} + R\psi_{x,tt}\,, \quad N_{6,x} + N_{2,y} = Pv_{0,tt} + R\psi_{y,tt}\,, \quad Q_{1,x} + Q_{2,y} = q + Pw_{,tt}\,, \\ &M_{1,x} + M_{6,y} - Q_1 = I\psi_{x,tt} + Ru_{0,tt}, \quad M_{6,x} + M_{2,y} - Q_2 = I\psi_{y,tt} + Rv_{0,tt} \end{aligned} \tag{2.3}$$

where P, R, and I are the normal, coupled-normal-rotatory, and rotatory inertia coefficients,

$$(P, R, I) = \sum_m \int_{z_m}^{z_{m+1}} (1, z, z^2)\, \varrho^{(m)}\, dz\,, \tag{2.4}$$

$\varrho^{(m)}$ being the material density of the m-th layer, q is the transversely distributed force, and N_i, Q_i, and M_i are the respective inplane and transverse stress and moment resultants defined by

$$(N_i, M_i) = \int_{-h/2}^{h/2} (1, z)\, \sigma_i\, dz \quad (i = 1, 2, 6), \qquad (Q_1, Q_2) = \int_{-h/2}^{h/2} (\sigma_5, \sigma_4)\, dz\,, \tag{2.5}$$

where h is the plate (laminate) thickness, and the so-called contracted subscript notation is employed to denote the stress components.

Assuming that the only plane of symmetry existing is in the plane of the plate, the theormoelastic constitutive relations for each layer (l) are taken to be monoclinic and bimodular as follows:

$$\begin{Bmatrix} \sigma_1 \\ \sigma_2 \\ \sigma_6 \end{Bmatrix} = \begin{bmatrix} Q_{11kl} & Q_{12kl} & 0 \\ Q_{12kl} & Q_{22kl} & 0 \\ 0 & 0 & Q_{66kl} \end{bmatrix} \begin{Bmatrix} \varepsilon_1 - \alpha_{1kl}T \\ \varepsilon_2 - \alpha_{2kl}T \\ \varepsilon_6 \end{Bmatrix}, \tag{2.6}$$

$$\begin{Bmatrix} \sigma_4 \\ \sigma_5 \end{Bmatrix} = \begin{bmatrix} C_{44kl} & C_{45kl} \\ C_{45kl} & C_{55kl} \end{bmatrix} \begin{Bmatrix} \varepsilon_4 \\ \varepsilon_5 \end{Bmatrix}. \tag{2.7}$$

Here T is the temperature change measured from the strain-free temperature, the C's are CAUCHY elastic shear stiffnesses, the Q's are plane-stress reduced stiffnesses, and the α's are thermal-expansion coefficients. The third subscript in Q_{ijkl} and C_{ijkl} (and second in α_{jkl}) refers to the bimodular characteristics: $k = 1$ denotes properties associated with fiber-direction tension, $k = 2$ denotes fiber-direction compression. Also, subscript l refers to the individual layer number, i.e., $l = 1$ and 2 for a two-layer laminate.

Substituting (2.2), (2.6) and (2.7) into (2.5), we obtain the constitutive equations for an arbitrarily laminated plate,

$$\begin{Bmatrix} \{N\} \\ \{M\} \end{Bmatrix} = \left[\begin{array}{c|c} [A] & [B] \\ \hline [B]^T & [D] \end{array}\right] \begin{Bmatrix} \{\varepsilon^0\} \\ \{\varkappa\} \end{Bmatrix} - \begin{Bmatrix} \{N^T\} \\ \{M^T\} \end{Bmatrix}, \tag{2.8}$$

$$\begin{Bmatrix} Q_2 \\ Q_1 \end{Bmatrix} = \begin{bmatrix} S_{44} & S_{45} \\ S_{45} & S_{55} \end{bmatrix} \begin{Bmatrix} \varepsilon_4 \\ \varepsilon_5 \end{Bmatrix}, \tag{2.9}$$

where

$$\begin{aligned} &\varepsilon_1^0 = u_{0,x}\,, \quad \varepsilon_2^0 = v_{0,y}\,, \quad \varepsilon_6^0 = u_{0,y} + v_{0,x}\,, \\ &\varkappa_1 = \psi_{x,x}\,, \quad \varkappa_2 = \psi_{y,y}\,, \quad \varkappa_6 = \psi_{x,y} + \psi_{y,x} \end{aligned} \tag{2.10}$$

and A_{ij}, B_{ij}, D_{ij}, S_{ij} are the respective inplane, bending-inplane coupling, bending or twisting, and thickness-shear stiffnesses defined as follows:

$$\begin{aligned} &(A_{ij}, B_{ij}, D_{ij}) = \int_{-h/2}^{h/2} (1, z, z^2)\, Q_{ijkl}\, dz \quad (i, j = 1, 2, 6)\,, \\ &S_{ij} = K_i K_j \int_{-h/2}^{h/2} Q_{ijkl}\, dz \quad (i, j = 4,5)\,. \end{aligned} \tag{2.11}$$

The stress and moment resultants, N_j^T and M_j^T, due to thermal loading, are defined by

$$(N_i^T, M_i^T) = \sum_m \int_{z_m}^{z_{m+1}} \sum_j^2 Q_{ijkm}\alpha_{ijkm}(T_0, zT_1)\,dz\,, \qquad (i = 1, 2, 6) \tag{2.12}$$

where the temperature change from a reference state is assumed to be linear through the thickness, consistent with plate theory:

$$T(x, y, z) = T_0(x, y) + zT_1(x, y)\,. \tag{2.13}$$

This completes the description of the field equations. Closed-form solutions to these equations are possible in the particular case of a cross-ply plate whose stiffness coefficients (A_{ij}, B_{ij}, and D_{ij}) with subscripts 16, 26 and 45 are zero, and whose edges are hinged flexurally but free to move in a direction normal to each edge. The temperature distribution and the normal-pressure loadings must be sinusoidal in order to obtain the closed-form solution (see [18–23]).

3. Finite-Element Formulation

The finite-element formulation of the equations presented in the previous section follows along the same lines as for ordinary (i.e., not bimodulus) plates (see REDDY [25]). The main difference lies in the computation of the stiffness coefficients, which depend on the stresses in the plate. We summarize the results in the following.

Consider a finite-element representation of the midplane of the plate by quadrilateral isoparametric elements. Over a typical element, R^e, the displacements (u_0, v_0, w, ψ_x, ψ_y) are approximated by expressions

$$u_0 = \sum_{i=1}^{n} u_i\varphi_i\,, \qquad v_0 = \sum_i^n v_i\varphi_i\,, \quad \text{etc}\,. \tag{3.1}$$

where $\varphi_i(x, y)$ $(i = 1, 2, \ldots, n)$ are interpolation functions associated with the linear $(n = 4)$ or quadratic $(n = 8 \text{ or } 9)$ rectangular element. Substituting (3.1) into the weak form associated with (2.3), (2.8), and (2.9), we obtain

$$([K] - \omega^2[M])\,\{\Delta\} = \{F\}\,. \tag{3.2}$$

The elements of the stiffness matrix $[K]$ and mass matrix $[M]$ are given by

$$\begin{aligned}
&K_{ij}^{11} = A_{11}G_{ij}^x + A_{66}G_{ij}^y\,, \quad K_{ij}^{12} = A_{12}G_{ij}^{xy} + A_{66}G_{ji}^{xy}\,, \quad K_{ij}^{14} = B_{11}G_{ij}^x + B_{66}G_{ij}^y\,,\\
&K_{ij}^{15} = B_{12}G_{ij}^{xy} + B_{66}G_{ji}^{xy}\,, \quad K_{ij}^{22} = A_{22}G_{ij}^y + A_{66}G_{ij}^x\,, \quad K_{ij}^{24} = B_{66}G_{ij}^{xy} + B_{12}G_{ji}^{xy}\,,\\
&K_{ij}^{25} = B_{66}G_{ij}^x + B_{22}G_{ij}^y\,, \quad K_{ij}^{33} = S_{55}G_{ij}^x + S_{44}G_{ij}^y\,, \quad K_{ij}^{34} = S_{55}G_{ij}^x + S_{44}G_{ij}^y\,, \quad K_{ij}^{35} = S_{44}G_{ij}^{y0}\,,\\
&K_{ij}^{44} = D_{11}G_{ij}^x + D_{66}G_{ij}^y + S_{55}G_{ij}^0\,, \quad K_{ij}^{45} = D_{12}G_{ij}^{xy} + D_{66}G_{ji}^{xy}\,,\\
&K_{ij}^{55} = D_{66}G_{ij}^x + D_{22}G_{ij}^y + S_{44}G_{ij}^0\,, \quad K_{ij}^{13} = K_{ij}^{23} = 0\,,
\end{aligned} \tag{3.3}$$

$$\begin{aligned}
&M_{ij}^{\alpha\alpha} = PG_{ij}^{00}\,, \qquad (\alpha = 1, 2,)\,,\\
&M_{ij}^{\alpha\beta} = RG_{ij}^{00}\,, \qquad (\alpha, \beta = 4{,}5; \alpha \neq \beta)\,, \qquad M_{ij}^{\alpha\alpha} = IG_{ij}^{00}\,, \qquad (\alpha = 4, 5)\,,
\end{aligned} \tag{3.4}$$

$$F_i^\alpha = \int_{R^e} f_\alpha\varphi_i\,dx\,dy\,, \qquad (\alpha = 1, 2, 3, 4, 5)\,, \tag{3.5}$$

$$G_{ij}^{\xi\eta} = \int_{R^e} \varphi_{i,\xi}\varphi_{j,\eta}\,dx\,dy\,, \qquad (\xi, \eta = 0, 1, 2)\,. \tag{3.6}$$

$$\begin{aligned}
&f_1 = N_{1,x}^T + N_{6,y}^T\,, \quad f_2 = N_{6,x}^T + N_{2,y}^T\,, \quad f_3 = q\,, \quad f_4 = M_{1,x}^T + M_{6,y}^T\,,\\
&f_5 = M_{6,x}^T + M_{2,y}^T\,.
\end{aligned} \tag{3.7}$$

In the present study the 4-node and 8-node rectangular isoparametric elements are employed.

4. Numerical Results and Discussion

Numerical results for deflections, neutral-surface locations, and vibration frequencies are presented for plates under various edge conditions and loadings. The discussion of the results is divided into three parts according to bending due to normal pressure, bending due to thermal loading, and free vibration. In all three parts, numerical results are presented first for plate problems which admit closed-form solutions. All of the finite-element results presented herein were obtained on an IBM 370/158 computer using the double-precision arithmetic.

Static Bending Due to Normal Pressure

To show the accuracy of the finite-element solutions, a comparison is made of the present results with the closed-form results obtained in [21]. Table 2 contains the neutral-surface locations, and the nondimensionalized transverse deflection of single-layer and two-layer (0°/90°), simply-supported (SS), rectangular plates (aramid-rubber) under sinusoidally (SSL) distributed transverse load, $q = q_0 \sin(\pi x/a)\sin(\pi y/b)$. As can be seen from Table 2, the agreement between the closed-form solution (CFS) and finite-element solution (FES) is extremely good. It should be noted that if the material is treated as an ordinary material, with material properties taken as the average of compressive and tensile properties, the deflections will be under-estimated significantly.

Table 1. Elastic properties for two tirecord-rubber, unidirectional, bimodulus composite materials

Property and Units	Aramid-Rubber (AR)		Polyester-Rubber (PR)	
	Tension ($k = 1$)	Compression ($k = 2$)	Tension ($k = 1$)	Compression ($k = 2$)
Longitudinal Young's modulus, GPa (E_{11})	3.58	0.0120	0.617	0.369
Transverse Young's modulus, GPa (E_{22})	0.00909	0.0120	0.00800	0.0106
Major Poisson's ratio, dimensionless[a] (ν_{12})	0.416	0.205	0.475	0.185
Longitudinal-transverse shear modulus, GPa[b] (G_{12})	0.00370	0.00370	0.00262	0.00267
Transverse-thickness shear modulus, GPa (G_{23})	0.00290	0.00499	0.00233	0.00475

[a]The minor Poisson's ratio is assumed to be given by the reciprocal relation.
[b]The longitudinal-thickness shear modulus is assumed to be equal to the longitudinal-transverse shear modulus.

Table 2. Neutral-surface locations and dimensionless center deflections for simply-supported thick rectangular plates of aramid-rubber ($b/h = 10$, shear correction factor $= 5/6$) subjected to sinusoidally distributed normal-pressure loading

	Single-layer					Two-layer (0°/90°)[2]				
	$Z_x = z_{nx}/h$		$\bar{w} = (w_0 E_{22}^C h^3/q_0 b^4)\,10^2$			$Z_x = z_{nx}/h$		$\bar{w} = (w_0 E_{22}^C h^3/q_0 b^4)\,10^2$		
b/a	CFS	FES	CFS	FES	Average[1]) (FES)	CFS	FES	CFS	FES	Average[1]) (FES)
0.6	0.4452	0.4452	0.456	0.456	0.1359	0.4427	0.4426	0.4388	0.434	0.2186
0.8	0.4440	0.4440	1.105	1.106	0.2546	0.4407	0.4407	1.054	1.047	0.3782
1.0	0.4420	0.4420	2.046	2.052	0.4239	0.4384	0.4384	1.957	1.946	0.5453
1.2	0.4394	0.4394	3.160	3.172	0.6561	0.4356	0.4357	3.043	3.029	0.6926
1.4	0.4363	0.4363	4.313	4.331	0.9664	0.4326	0.4326	4.185	4.167	0.8078
1.6	0.4328	0.4329	5.406	5.434	1.3724	0.4292	0.4293	5.282	5.261	0.8918
1.8	0.4292	0.4294	6.390	6.426	1.8943	0.4257	0.4257	6.277	6.254	0.9511
2.0	0.4253	0.4254	7.250	7.293	—	0.4219	0.4219	7.151	7.127	—

1) Results obtained by using the average of the compressive and tensile properties listed in Table 1.
2) Neutral-surface location Z_y is not listed here (see [21]).

The aramid-rubber plates, in both the single- and cross-ply cases, have noticeably larger values of Z_x than the polyester-rubber ones, due to the more pronounced bimodulus effect in the fiber-direction Young's modulus of the aramid-rubber. There are only very slight differences in Z_x and deflection in going from a single-layer plate to a crossply laminate. This is in considerable contrast to ordinary materials.

Having validated the finite-element model, the model is then used for other combinations of loading and boundary conditions that are not amenable to closed-form solutions. Figure 1 shows plots of the nondimensionalized

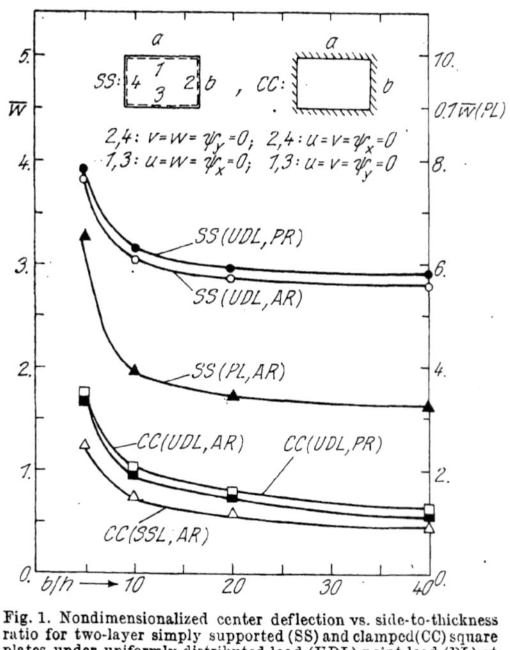

Fig. 1. Nondimensionalized center deflection vs. side-to-thickness ratio for two-layer simply supported (SS) and clamped (CC) square plates under uniformly distributed load (UDL) point load (PL) at the center

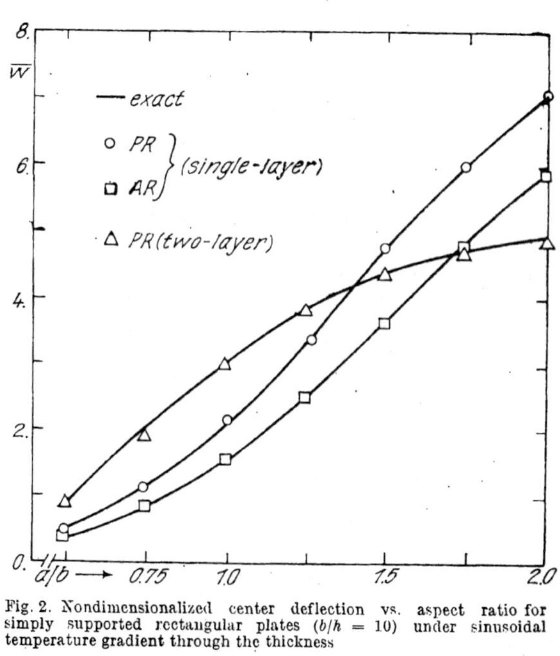

Fig. 2. Nondimensionalized center deflection vs. aspect ratio for simply supported rectangular plates ($b/h = 10$) under sinusoidal temperature gradient through the thickness

deflection versus the side-to-thickness ratio for two-layer (0°/90°) simply-supported and clamped (CC) square plates under various loadings. The effect of the thickness shear on the deflections is apparent from the graphs.

Static Bending Due to Thermal Loading

Next, numerical results are presented for static bending of plates under thermal gradients only (i.e., $\overline{T}_0 = 0$). The same orthotropic bimodulus materials as listed in Table 1 are used. Since no measured values of thermal-expansion coefficients of these materials are available, the following ratios of the thermal-expansion coefficients are used:

$$\alpha_1^t/\alpha_1^C = 0.5; \qquad \alpha_2^t/\alpha_2^C = 1.0; \qquad \alpha_1^t/\alpha_2^t = 0.1 \,. \tag{4.3}$$

To show the accuracy of the finite-element solutions, a comparison is made of the present results with the closed-form results obtained in [22]. Table 3 contains the neutral-surface locations, and the nondimensionalized transverse deflection of single-layer and two-layer (0°/90°), simply-supported (SS), rectangular plates (polyester-rubber) under sinusoidally (SST) distributed temperature, $T = \overline{T}_1 z \sin(\pi x/a) \sin(\pi y/b)$. As can be seen from Table3, the agreement between the closed-form solution (CFS) and finite-element solution (FES) is extremely good (see also Fig. 2).

Table 3. Neutral-surface locations and dimensionless center deflections for simply-supported thick ($b/h = 10$, $K_1 = K_2 = 5/6$) rectangular plates of polyester-rubber subjected to sinusoidal temperature gradient, $T = z\overline{T}_1 \sin(\pi x/a) \times \sin(\pi y/b)$

	single-layer				two-layer *)			
	Z_x (= z_x/h)		$\overline{w}$**)		Z_x (= z_x/h)		$\overline{w}$	
a/b	CFS	FES	CFS	FES	CFS	FES	CFS	FES
0.50	0.1031	0.1030	0.482	0.482	0.2599	0.2541	0.928	0.894
0.75	0.1184	0.1183	1.157	0.158	0.2554	0.2436	0.199	0.187
1.00	0.1308	0.1308	2.160	2.161	0.2398	0.2294	0.309	0.296
1.25	0.1360	0.1360	3.410	3.409	0.2119	0.2035	0.392	0.382
1.50	0.1332	0.1331	4.737	4.736	0.1734	0.1679	0.443	0.436
1.75	0.1234	0.1233	5.975	5.970	0.1284	0.1252	0.472	0.466
2.0	0.1078	0.1078	7.024	7.017	0.0815	0.0804	0.489	0.483

*) Neutral-surface location Z_y is not listed here (see [22]).
**) $\overline{w} = (wh/\alpha_1^t \overline{T}_1 b^2)\, 10$.

Next, finite-element results are presented for other combinations of loading and boundary conditions that are not amenable to closed-form solutions. Figure 3 shows plots of the nondimensionalized deflection versus the side-to-thickness ratio for two-layer (0°/90°) and one-layer simply-supported (SS) square plates (polyester-rubber) under uniform temperature gradient through the thickness. The effect of the thickness shear on the deflections is apparent from the graphs.

Natural Vibration Results

Before we present the numerical results of free vibration, we discuss the effect of bimodulus action on plate stiffness in different portions of a vibration cycle. First consider a single-layer, bimodulus material plate at the two extremes of its deflection (Fig. 4). During the first half cycle (see Fig. 4A) the top surface is in compression and the bottom

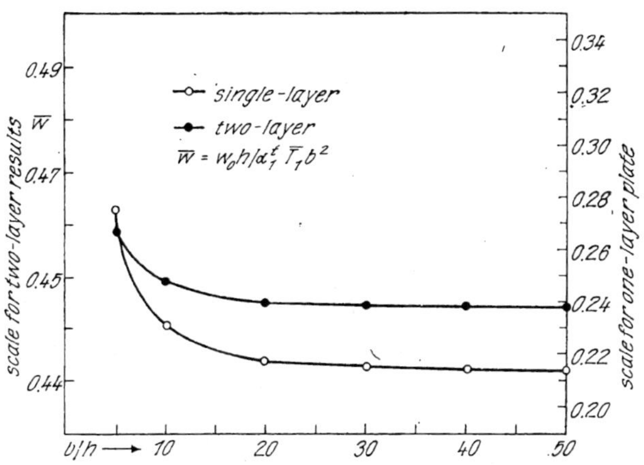

Fig. 3. Nondimensionalized center deflection vs. side-to-thickness ratio for two-layer simply supported (SS) square plates under uniformly distributed temperature gradient through the thickness (instead of 0.44 on the right hand vertical axis read 0.43)

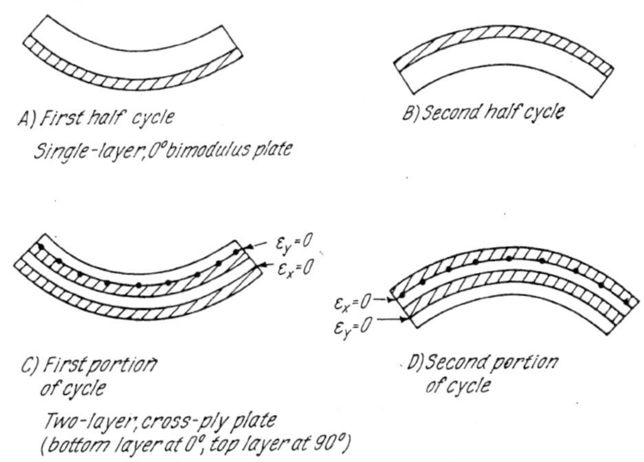

Fig. 4. Bimodulus action in single-layer and cross-ply plates in the fundamental mode of vibration. Cross-hatched regions denote tension in the respective fiber directions

in tension, thus causing the neutral surface for ε_x to be positive ($z_{nx} > 0$), i.e., a certain distance below the plate midplane. During the second half cycle (Fig. 4B) the top surface is in tension and the bottom is in compression, just opposite to that encountered in the first half cycle. However, the magnitude of z_{nx} is the same as in the first half cycle. Thus, it can be concluded that the effective stiffness (and thus the frequency) associated with the second half cycle is identical to that of the first half cycle and either modal shape, Figs. 4A or 4B, will give the same computational result for the natural frequencies.

Now consider a two-ply laminate with the bottom layer (layer $l = 1$) oriented at 0° and the top layer ($l = 2$) at 90° (see Fig. 4). Initially, as shown in Fig. 4C, the neutral surface for ε_x falls below the interface, within the 0° layer, while the neutral surface for ε_y falls above the interface, completely within the 90° layer. In the latter portion of the cycle, Fig. 4D, the ε_x neutral surface falls outside of the 0° layer, and the ε_y neutral surface falls outside of the 90° layer. Therefore, compressive properties are used for the entire 0° layer, and tensile ones for the 90° layer. Thus, for a two-layer cross-ply laminate the plate stiffnesses acting in the two portions of a cycle are different and the associated frequencies are also different (except for a square plate). Denoting the frequencies associated with the two portions of a vibration cycle by ω_1 and ω_2, we can easily show that the average frequency (ω) over the entire cycle is

$$\omega^{-1} = (1/2)\,(\omega_1^{-1} + \omega_2^{-1})\,. \tag{4.4}$$

Thus, the computational procedure required for a cross-ply plate is to calculate ω_1 and ω_2 associated with modal shapes shown in Figs. 4C and 4D, respectively, and then to apply equation (4.4).

The question arises as to the possibility of a discontinuity in energy at the instant of time between the two portions of a cycle. However, it was shown in [23] that there is no such discontinuity.

Numerical results are presented for relatively thick plates ($b/h = 10$) of aramid-rubber. Plate aspect ratio affects the fundamental frequency much more than it does the neutral-surface portion (see Tables 4—5). For the cross-ply case, only $Z_x^{(1)}$ is close to $Z_y^{(2)}$. In Tab. 5 the good agreement between the CFS and FES results is clearly shown.

Since there are very drastic changes in neutral-surface locations (for example, z_{nx} for aramid-rubber goes from approximately $0.4h$ to $0.03h$) from one cycle portion to the other, there is a question that may arise regarding a possible transient action. However, the neutral surfaces are just boundaries between the tensile and compressive regions (analogous to the elastic-plastic boundary in elastoplastic problems). Thus, they have no mass and there is no transient action.

Table 4. Dimensionless fiber-direction neutral-surface locations for single-layer (0°) and two-layer (0°/90°) plates of polyester-rubber material ($b/h = 10$)

	single-layer (0°);		polyester-rubber		two-layer (0°/90°); polyester-rubber[1]							
	$\omega_0 b^2(P/E_{22}^C h^3)^{1/2}$		$Z_x = z_{nx}/h$		$Z_x^{(1)}$		$-Z_y^{(1)}$		$-Z_x^{(2)}$		$Z_y^{(2)}$	
a/b	CFS	FES	CFS	FES	CFS	FES	CFS	FES	CFS	FES	CFS	FES
0.5	25.134	23.136	0.3089	0.3083	0.3687	0.3691	0.1335	0.1295	0.0830	0.0825	0.3569	0.357
0.7	15.058	14.421	0.3072	0.3071	0.3664	0.3663	0.1119	0.1113	0.0868	0.0868	0.3603	0.360
1.0	8.668	8.648	0.3056	0.3049	0.3632	0.3633	0.0960	0.0960	0.0959	0.0959	0.3631	0.363
1.4	5.421	5.533	0.3011	0.3013	0.3589	0.3596	0.0870	0.0870	0.1115	0.112	0.3648	0.365
2.0	3.777	3.918	0.2945	0.2950	0.3514	0.3513	0.0817	0.0817	0.1389	0.139	0.3660	0.366

[1] $Z_x^{(i)}$ denotes $z_{nx}h$ for the i-th ($i = 1, 2$) portion of the cycle.

Table 5. Dimensionless fundamental frequencies in the first partial cycle, second partial cycle, and complete cycle of motion for two-layer, cross-ply plates having $b/h = 10$ by closed form and finite element methods

	$\omega_1 b^2(P/E_{22}^C h^3)^{1/2}$		$\omega_2 b^2(P/E_{22}^C h^3)^{1/2}$		$\omega b^2(P/E_{22}^C h^3)^{1/2}$	
a/b	CFS	FES	CFS	FES	CFS	FES
Aramid-Rubber						
0.5	19.38	20.23	13.88	14.55	16.18	16.93
0.7	11.60	12.17	9.353	9.807	10.35	10.86
1.0	7.038	7.386	7.038	7.364	7.038	7.375
1.4	4.838	5.045	6.037	6.356	5.371	5.625
2.0	3.712	3.909	5.551	5.821	4.449	4.677
Polyester-Rubber						
0.5	19.12	19.81	15.95	16.61	17.39	18.07
0.7	11.43	11.92	10.04	10.45	10.69	11.14
1.0	7.084	7.406	7.085	7.394	7.085	7.400
1.4	5.164	5.407	5.928	6.193	5.520	5.773
2.0	4.310	4.518	5.435	5.688	4.807	5.036

5. Summary and Conclusions

Both closed-form and finite-element solutions are presented for static bending and free vibrations of single-layer and two-layer cross-ply plates of bimodulus composite materials. In the case of simply supported plates under sinusoidal temperature distribution, normal-pressure loading, and free vibration, the closed-form and the finite-element solutions for neutral-surface locations, center deflections, and fundamental frequencies are in excellent agreement. The finite-element method is then used to obtain solutions for simply-supported plates under uniform loading, for which no closed-form solution can be obtained for the bimodulus case.

The research reported here has recently been extended to mechanically loaded thin cylindrical shells [26], free vibration of thin and thick cylindrical shells [27], and thermally and mechanically loaded thick cylindrical shells [28].

Acknowledgements

The authors are grateful to the Office of Naval Research, Structural Mechanics Program for financial support through Contract N00014-78-C-0647. The authors also acknowledge the computational assistance of W. C. CHAO, Y. S. HSU, and V. S. REDDY.

References

1 CLARK, S. J., The plane elastic characteristics of cord-rubber laminates, Textile Research J. **33**, (1963), pp. 295–313.
2 PATEL, H. P.; TURNER, J. L.; WALTER, J. D., Radial tire cord-rubber composites, Rubber Chem. and Tech. **49**, (1976), pp. 1095–1110.
3 AMBARTSUMYAN, S. A., The basic equations and relations of the different-modulus theory of elasticity of an anisotropic body, Mechanics of Solids **4** (1969), pp. 48–56.
4 BERT, C. W., Recent advances in mathematical modeling of the mechanics of bimodulus, fiber-reinforced composite materials, Proc. 15th Annual Meeting, Society of Engineering Science, Gainesville, FL, Dec. 4–6 (1978), pp. 101–106.
5 BERT, C. W., Models for fibrous composites with different properties in tension and compression, J. Eng. Matls. and Tech., Trans. ASME **99H**, (1977), pp. 344–349.
6 SHAPIRO, G. S., Deformation of bodies with different tensile and compressive strengths [stiffnesses], Mechanics of Solids **1**, No. 2 (1966), pp. 85–86.
7 KAMIYA, N., Large deflection of a different modulus circular plate, Journal of Engineering Materials and Technology, Trans. ASME, **97H**, No. 1 (Jan. 1975), pp. 52–56.
8 KAMIYA, N., An energy method applied to large elastic deflection of a thin plate of bimodulus material, Journal of Structural Mechanics, **3**, No. 3 (1976), pp. 317–329.
9 KAMIYA, N., Transverse shear effect in a bimodulus plate, Nuclear Engineering and Design, **32**, No. 3 (July 1975), pp. 351–357.
10 JONES, R. M.; MORGAN, H. S., Bending and extension of cross-ply laminates with different moduli in tension and compression, AIAA/ASME/SAE 17th Structures, Structural Dynamics, and Materials Conference, King of Prussia, PA, May 5–7 (1976), pp. 158–167; Computers and Structures, **11**, No. 3, (Mar. 1980), pp. 181–190.
11 BERT, C. W.; KINCANNON, S. K., Bending-extensional coupling in elliptic plates of orthotropic bimodulus material, Developments in Mechanics, **10** (Proc. 16th Midwestern Mechanics Conference), Kansas State University, Manhattan, KS (Sept. 1979), pp. 7–11.
12 KINCANNON, S. K.; BERT, C. W.; REDDY, V. S., Cross-ply elliptic plates of bimodulus material, Journal of the Structural Division, Proc. ASCE, Vol. **106**, No. ST7, (1980), pp. 1437–1449.
13 AMBARTSUMYAN, S. A., The equations of temperature stresses of different-modulus materials, Mechanics of Solids, **3**, No. 5, (1968), pp. 58–69.
14 AMBARTSUMYAN, S. A., Equations of theory of thermal stresses in double-modulus materials, Thermoinelasticity (Proc., IUTAM Symposium, E. Kilbride, Scotland, June 25–28, 1968), BOLEY, B. A., ed., (1970), pp. 17–32, Springer-Verlag, Wien.
15 KAMIYA, N., Thermal stress in a bimodulus thin plate, Bulletin de l'academie Polonaise des Sciences, Série des sciences techniques, **24**, (1976), pp. 365–372.
16 KAMIYA, N., Energy formulae of bimodulus material in thermal field, Fibre Science and Technology, **11**, (1978), pp. 229–235.
17 KAMIYA, N., Bimodulus thermoelasticity considering temperature-dependent material properties, Mechanics of Bimodulus Materials, BERT, C. W., ed., AMD Vol. **33**, ASME, (Dec. 1979), pp. 29–37.
18 REDDY, J. N.; BERT, C. W., Analyses of plates constructed of fiber-reinforced bimodulus materials, Mechanics of Bimodulus Materials, BERT, C. W. ed., AMD Vol. **33**, ASME, NY, (Dec. 1979), pp. 67–83.
19 REDDY, J. N.; CHAO, W. C., Finite-element analysis of laminated bimodulus composite-material plates, Computers and Structures, **12**, (1980), pp. 245–251.
20 BERT, C. W.; REDDY, V. S.; KINCANNON, S. K., Deflection of thin rectangular plates of cross-plied bimodulus material, Journal of Structural Mechanics, **8**, No. 4, (1980), pp. 347–364.
21 BERT, C. W.; REDDY, J. N.; REDDY, V. S.; CHAO, W. C., Bending of thick rectangular plates laminated of bimodulus composite materials, AIAA Journal **19**, No. 10 (1981), pp. 1342–1349.
22 REDDY, J. N.; BERT, C. W.; HSU, Y. S.; REDDY, V. S., Thermal bending of thick rectangular plates of bimodulus composites materials, Journal of Mechanical Engineering Science, **22**, No. 4 (1980), pp. 297–304.
23 BERT, C. W.; REDDY, J. N.; CHAO, W. C.; REDDY, V. S., Vibration of thick rectangular plates of bimodulus composite material, Journal of Applied Mechanics, **48** (1981), pp. 371–376.
24 WHITNEY, J. M.; PAGANO, N., Shear deformation in heterogeneous anisotropic plates, Journal of Applied Mechanics, **37**, (1970), pp. 1031–1036.
25 REDDY, J. N., A penalty-plate bending element for the analysis of laminated anisotropic composite plates, International Journal for Numerical Methods in Engineering, **15**, No. 8, (Aug. 1980), pp. 1187–1206.
26 BERT, C. W.; REDDY, V. S., Cylindrical shells of bimodulus material, ASCE Preprint 80-623, presented at the ASCE National Convention and Exposition, Hollywood, FL, (Oct. 27–31, 1980).
27 BERT, C. W.; KUMAR, M., Vibration of cylindrical shells of bimodulus composite materials, AIAA/ASME/ASCE/AHS 22nd Structures, Structural Dynamics and Materials Conference, Atlanta, GA, (Apr. 6–8, 1981).
28 HSU, Y. S.; REDDY, J. N.; BERT, C. W., Thermoelasticity of circular cylindrical shells laminated of bimodulus composite materials, Journal of Thermal Stresses, **4** (1981), pp. 155–177.

Eingereicht am 30. 4. 1981

Anschrift: Prof. J. N. REDDY, Engineering Science and Mechanics, Virginia Polytechnic Institute and State University, Blacksburg, Virginia 24061, USA
Prof. C. W. BERT, School of Aerospace Mechanical and Nuclear Engineering, University of Oklahoma, Norman, Oklahoma 73016, USA

ZAMM 62, 221 – 228 (1982)

Y. TANIGAWA / Y. TAKEUTI

Coupled Thermoelastic Problem in Three-Dimensions

Zur Behandlung dreidimensionaler gekoppelter thermo-elastischer Probleme wird eine neue Technik vorgeschlagen. Man führt eine additive harmonische Funktion in die beschreibenden Beziehungen ein. Danach kann man in thermische und elastische Differentialgleichung trennen. Wenn man diese Technik ausnutzt, ist man in der Lage, die exakte Lösung für den gesamten Zeitbereich des dreidimensionalen transienten Wärmespannungsproblems eines Kreiszylinders zu erhalten. Zum Vergleich mit klassischen Resultaten werden mehrere numerische Ergebnisse für Temperatur- und Spannungsverteilungen angegeben.

A new technique is proposed to the coupled thermoelastic problems in three-dimensions. In the basic relations we introduce an additive harmonic function and then it is possible to separate both thermal and elastic fundamental differential equations Taking advantage of this technique we will be able to obtain the exact solution for the whole time range in the three-dimen sional transient thermal stress problem of a circular cylinder. Several numerical results in temperature and stress distribution are given for the sake of comparison to the classical results.

Для решения трехмерных связанных термо-упругих задач предлагается новый метод. В основные уравнения вводится аддитивная гармоническая функция. Впоследствие этого разделяют в термическое и упругое дифференциальные управнения. С помощью этого метода можно получить точное решение нестационарной трехмерной задачи температурного напряжения круглого цилиндра на весь промежуток времени. Для сравнения с классическими результатами дано несколько численных результатов гля распределения температуры и напряжения.

1. Introduction

The coupled thermoelastic problem which employs a corrected heat conduction equation with the coupling term of thermal and displacement field requires the simultaneous determination of temperature and deformation. In the case of steep temperature gradient in engineering problems, it is important to consider the dynamic effect (the inertia term) and the coupling effect in the fundamental equations. From our previous result [1], however, inertia effects may disappear in pure thermal stress problems in contrast to the coupling effects results in a visible lag in the stress distribution for the usual case.

Since the inertia term gives very small effect for the realistic materials, it is only necessary to consider the coupling term in the governing equations. The present study devoted to the coupled thermoelastic problem in three-dimension has been developed by means of a new technique introducing an additive harmonic function. Thus it may be possible to separate both thermal and elastic fundamental differential equations. After this treatment we use successive integral transforms to obtain the solution. As an illustration, we apply the above method to the general three-dimensional thermal stress problem of an infinite circular cylinder under partial heating of circumferential and axial directions. Numerial calculation was carried out for the temperature and stress distributions using the realistic values of coupling parameter. Main advantage of this method is to obtain the exact solution for the whole time range.

2. Analysis

We shall consider the three-dimensional coupled thermoelastic problems governed by the three-dimensional orthogonal curvilinear systems. For the coupled thermoelastic problems, it is well-known that the heat conduction equation and the field equations of motion should be solved simultaneously so as to satisfy both the thermal and mechanical boundary conditions. Temperature field in consideration should be determined from the corrected heat conduction equation shown as

$$T_{,t} = \varkappa \Delta T - \frac{\tilde{T} \alpha E/(1 - 2\nu)}{C_v \tilde{\varrho}} e_{,t} \tag{1}$$

in which the comma denotes partial differentiation with respect to a variable, Δ is Laplacian operator, T is the temperature change, $\varkappa$ is the thermal diffusivity, t is time, e is dilatation or volumetric strain, $\tilde{T}$ is an absolute temperature for which the material is stress free, α is the coefficient of linear thermal expansion, E is Young's modulus of elasticity, C_v is the specific heat at constant volume and $\tilde{\varrho}$ is density.

The above expression (1) is valid for any three-dimensional coordinate system such as Cartesian coordinate (x, y, z), cylindrical coordinate (r, θ, z) and spherical coordinate (r, θ, φ) and so on. And the second term of right hand side of eq. (1) is the so-called thermocoupling term which means the heat energy consumed to the deformation of the body.

On the other side, the field equations of motion are given by the following relation

$$\mu_0 \nabla^2 \boldsymbol{u} + (\lambda_0 + \mu_0) \operatorname{grad} \operatorname{div} \boldsymbol{u} = (3\lambda_0 + 2\mu_0) \alpha \operatorname{grad} T \tag{2}$$

in which $\boldsymbol{u}$ is displacement vector, λ_0 and μ_0 are the Lamé's constants. As was cited in our previous paper [1], inertia term gives very small effect for the coupled thermoelastic problem when the realistic materials are taken into consideration, which means that we should consider only the thermocoupling term in the governing equation. Then, we shall disregard the inertia terms on field equations of motion as given by eq. (2).

For the illustrative purpose, we now consider the domain governed by the cylindrical coordinate system (r, θ, z) available for the solid or hollow circular cylinder. Then, eqs. (1) and (2) become to a following dimensionless form

$$\overline{T}_{,\tau} = \overline{\Delta}\overline{T} - (1-\nu)(1+\nu)^{-1}\delta\bar{e}_{,\tau}, \tag{3}$$

$$2(1-\nu)(1-2\nu)^{-1}\bar{e}_{,\varrho} + \varrho^{-2}\bar{u}_{,\theta\theta} + \bar{u}_{,\zeta\zeta} - \varrho^{-2}(\varrho\bar{v})_{,\varrho\theta} - \bar{w}_{,\varrho\zeta} = 2(1+\nu)(1-2\nu)^{-1}\overline{T}_{,\varrho}, \tag{4}$$

$$2(1-\nu)(1-2\nu)^{-1}\bar{e}_{,\theta} - \varrho(\bar{u}\varrho^{-1})_{,\varrho\theta} + \varrho\{\varrho^{-1}(\varrho\bar{v})_{,\varrho}\}_{,\varrho} + \varrho\bar{v}_{,\zeta\zeta} - \bar{w}_{,\theta\zeta} = 2(1+\nu)(1-2\nu)^{-1}\overline{T}_{,\theta}, \tag{5}$$

$$2(1-\nu)(1-2\nu)^{-1}\bar{e}_{,\zeta} - \varrho^{-1}(\varrho\bar{u})_{,\varrho\zeta} - \varrho^{-1}\bar{v}_{,\theta\zeta} + \varrho^{-1}(\varrho\bar{w}_{,\varrho})_{,\varrho} + \varrho^{-2}\bar{w}_{,\theta\theta} = 2(1+\nu)(1-2\nu)^{-1}\overline{T}_{,\zeta}, \tag{6}$$

where Laplacian operator and dilatation in dimensionless form are given as follows:

$$\overline{\Delta} = \partial^2/\partial\varrho^2 + \varrho^{-1}\partial/\partial\varrho + \varrho^{-2}\partial^2/\partial\theta^2 + \partial^2/\partial\zeta^2, \qquad \bar{e} = \bar{u}_{,\varrho} + \bar{u}\varrho^{-1} + \varrho^{-1}\bar{v}_{,\theta} + \bar{w}_{,\zeta}. \tag{7), (8}$$

In the above expressions (3)∼(8), the following dimensionless quantities are introduced

$$\varrho = r/a, \quad \zeta = z/a, \quad \tau = \varkappa t/a^2, \quad \overline{T} = T/T_0, \quad (\bar{u}, \bar{v}, \bar{w}) = (u, v, w)/\alpha T_0 a, \tag{9}$$

where r, θ and z are the radial, tangential and axial coordinate, and u, v and w are the corresponding displacement, respectively, and T_0 is a constant temperature prescribed by some heat source, a is a radius of a cylinder and δ appeared in eq. (3) is the coupling parameter

$$\delta = (1+\nu)\{(1-\nu)(1-2\nu)\}^{-1}\tilde{T}\alpha^2 E/(C_r\tilde{\varrho}). \tag{10}$$

On the other hand, the stress-strain relations expressed by dimensionless forms are

$$\begin{aligned}
\bar{\sigma}_{rr} &= (1+\nu)^{-1}\{\bar{u}_{,\varrho} + \nu\bar{e}/(1-2\nu)\} - \overline{T}/(1-2\nu), \\
\bar{\sigma}_{\theta\theta} &= (1+\nu)^{-1}\{\bar{u}\varrho^{-1} + \varrho^{-1}\bar{v}_{,\theta} + \nu\bar{e}/(1-2\nu)\} - \overline{T}/(1-2\nu), \\
\bar{\sigma}_{zz} &= (1+\nu)^{-1}\{\bar{w}_{,\zeta} + \nu\bar{e}/(1-2\nu)\} - \overline{T}/(1-2\nu), \\
\bar{\sigma}_{r\theta} &= (\bar{v}_{,\varrho} + \varrho^{-1}\bar{u}_{,\theta} - \bar{v}\varrho^{-1})/\{2(1+\nu)\}, \qquad \bar{\sigma}_{\theta z} = (\varrho^{-1}\bar{w}_{,\theta} + \bar{v}_{,\zeta})/\{2(1+\nu)\}, \\
\bar{\sigma}_{zr} &= (\bar{u}_{,\zeta} + \bar{w}_{,\varrho})/\{2(1+\nu)\},
\end{aligned} \tag{11}$$

in which $\bar{\sigma}_{ij}$ is denoted by $\bar{\sigma}_{ij} = \sigma_{ij}/\alpha E T_0$.

In order to solve the corrected heat conduction equation (3) and the governing equations (4)—(6) with respect to the associated stress fields simultaneously, we use a relationship connecting the temperature change $\overline{T}$ with dilatation $\bar{e}$:

$$\overline{\Delta}\bar{e} = (1+\nu)(1-\nu)^{-1}\overline{\Delta}\overline{T}. \tag{12}$$

If we now introduce a harmonic function Φ,

$$\overline{\Delta}\Phi = 0, \tag{13}$$

and substitute this notion into eq. (12), then we have a following relation between temperature change and dilatation

$$\bar{e} = (1+\nu)(1-\nu)^{-1}\{\overline{T} + \Phi\}. \tag{14}$$

It can be shown that the relations (13) and (14) are also valid for the other coordinate systems such as Cartesian coordinates (x, y, z) or spherical coordinates (r, θ, φ) by using the Laplacian operator $\overline{\Delta}$ evaluated in each coordinate system.

Substituting eq. (14) into eqs. (3)—(6), the governing equations for the three-dimensional coupled thermoelastic problem are shown in terms of the harmonic function Φ

$$\overline{\Delta}\overline{T} = (1+\delta)\overline{T}_{,\tau} + \delta\Phi_{,\tau}, \qquad 2(1+\nu)(1-2\nu)^{-1}\Phi_{,\varrho} + \varrho^{-2}\bar{u}_{,\theta\theta} + \bar{u}_{,\zeta\zeta} - \varrho^{-2}(\varrho\bar{v})'_{,\varrho\theta} - \bar{w}_{,\varrho\zeta} = 0, \tag{15), (16}$$

$$2(1+\nu)(1-2\nu)^{-1}\Phi_{,\theta} - \varrho(\bar{u}\varrho^{-1})_{,\varrho\theta} + \varrho\{\varrho^{-1}(\varrho\bar{v})_{,\varrho}\}_{,\varrho} + \varrho\bar{v}_{,\zeta\zeta} - \bar{w}_{,\theta\zeta} = 0, \tag{17}$$

$$2(1+\nu)(1-2\nu)^{-1}\Phi_{,\zeta} - \varrho^{-1}(\varrho\bar{u})_{,\varrho\zeta} - \varrho^{-1}\bar{v}_{,\theta\zeta} + \varrho^{-1}(\varrho\bar{w}_{,\zeta})_{,\varrho} + \varrho^{-2}\bar{w}_{,\theta\theta} = 0. \tag{18}$$

From these replaced fundamental equations, it can be seen that the coupled heat conduction equation and the corresponding stress fields can be treated individually with aid of the harmonic function Φ, which means that the temperature field and the associated stress fields can be treated introducing an additive harmonic function Φ in the same manner as the usual uncoupled thermoelastic problems. In addition to the above fundamental relation (13) and (15)—(18), we have the following relation from eqs. (8) and (14):

$$\bar{u}_{,\varrho} + \bar{u}\varrho^{-1} + \varrho^{-1}\bar{v}_{,\theta} + \bar{w}_{,\zeta} = (1+\nu)(1-\nu)^{-1}\{\overline{T} + \Phi\}. \tag{19}$$

In order to develop the analysis for practical problems, we now assume that the infinitely long solid cylinder is suddenly heated by surrounding medium of which the temperature varies both in lateral and tangential directions

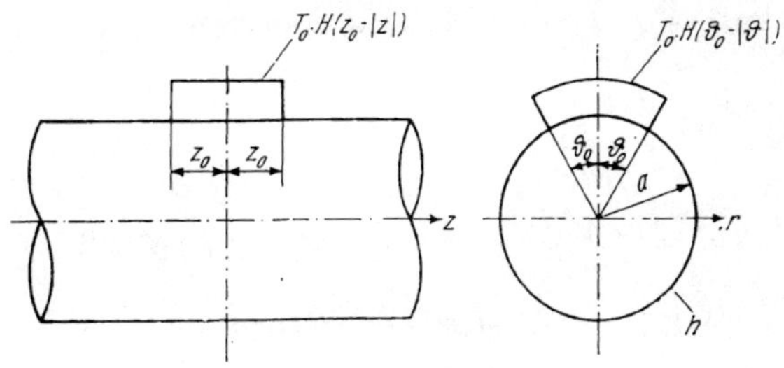

Fig. 1. Coordinate system and thermal boundary condition of a solid cylinder

as shown in Fig. 1. Then, the temperature change of the cylinder becomes to be three-dimensional distribution. We assume that the temperature distribution of surrounding medium is given by

$$T_s = T_0 f(\theta)\, g(\zeta)\,. \tag{20}$$

Then, the initial and boundary conditions for the heat conduction problem are taken in dimensionless form as

$$\tau = 0; \quad \overline{T} = 0\,, \quad \Phi = 0\,, \quad \varrho = 1; \quad \overline{T}_{,\varrho} + H_a \overline{T} = H_a f(\theta)\, g(\zeta); \quad H_a \equiv ah\,, \tag{21), (22}$$

in which h is relative heat transfer coefficient on the surface. The above initial condition for Φ can be obtained from eq. (14) by introducing an assumption that both the initial temperature and initial strain can be regarded as zero. On the other side, if we assume that the surface is traction-free, the mechanical boundary conditions can be taken as

$$\varrho = 1; \quad \bar{\sigma}_{rr} = 0\,, \quad \bar{\sigma}_{r\theta} = 0\,, \quad \bar{\sigma}_{rz} = 0\,. \tag{23}$$

For the sake of the simplification of analysis, we shall introduce the following symmetrical condition

$$f(\theta) = f(-\theta)\,, \qquad g(\zeta) = g(-\zeta)\,, \tag{24}$$

and develop the analytical formulation by using the successive integral transformation and method of separation of variables. Denoting Laplace transform of F(τ) by $F^*(p)$, and Fourier-cosine transform of $G(\zeta)$ by $\hat{G}(q)$, then the governing equation for Φ in transformed domain becomes

$$\hat{\Phi}^*_{,\varrho\varrho} + \varrho^{-1}\hat{\Phi}^*_{,\varrho} + \varrho^{-2}\hat{\Phi}^*_{,\theta\theta} - q^2\hat{\Phi}^* = 0\,. \tag{25}$$

Then the fundamental solution available for the solid cylinder is shown as

$$\Phi^* = \sqrt{2/\pi} \int_0^\infty \sum_{n=0}^\infty A_n I_n(q\varrho) \cos n\theta \cos q\zeta \, dq\,. \tag{26}$$

Similarly, we can apply the successive transform to the heat conduction equation (15) taking into account the initial condition (21), then the basic equation for $\overline{T}$ in the transformed domain takes the following form

$$\hat{\overline{T}}^*_{,\varrho\varrho} + \varrho^{-1}\,\hat{\overline{T}}^*_{,\varrho} + \varrho^{-2}\,\hat{\overline{T}}^*_{,\theta\theta} - \mu^2\hat{\overline{T}}^* = \delta p \hat{\Phi}^*\,, \tag{27}$$

$$\mu^2 = q^2 + (1 + \delta)\, p\,. \tag{28}$$

Eq. (27) can be solved with aid of eq. (26) and the fundamental solution $\overline{T}^*$ is

$$\overline{T}^* = \sqrt{2/\pi} \int_0^\infty \sum_{n=0}^\infty \{B_n I_n(\mu\varrho) - \delta(1 + \delta)^{-1} A_n I_n(q\varrho)\} \cos n\theta \cos q\zeta \, dq\,, \tag{29}$$

in which A_n and B_n are the unknown constants to be determined from the boundary conditions.

We shall develop the analysis for the stress fields. In the case of three-dimensional problem, displacements $\bar{u}$, $\bar{v}$ and $\bar{w}$ should be determined by solving the field equations of motion given by eqs. (16)—(18) and (19). Eliminating the term $\bar{w}$ from eqs. (16) and (19) and applying the same manner to eqs. (17) and (19), then the following two relations are obtained

$$\bar{\Delta}\, \bar{u} - \varrho^{-2}\, \bar{u} - 2\varrho^{-2}\bar{v}_{,\theta} = (1 + \nu)\,(1 - \nu)^{-1}\, \{\overline{T} - \Phi/(1 - 2\nu)\}_{,\varrho}\,, \tag{30}$$

$$\Delta\, \bar{v} - \varrho^{-2}\bar{v} + 2\varrho^{-2}\bar{u}_{,\theta} = (1 + \nu)\,(1 - \nu)^{-1}\, \varrho^{-1}\, \{\overline{T} - \Phi/(1 - 2\nu)\}_{,\theta}\,. \tag{31}$$

From the above two relations, we have a following relation between $\bar{u}$ and $\bar{v}$:

$$\bar{\Delta}(\varrho^{-1}\bar{u}_{,\theta}) = \bar{\Delta}\{\varrho^{-1}(\varrho\bar{v})_{,\varrho}\}\,. \tag{32}$$

If we introduce another harmonic function Ψ such that

$$\bar{\Delta}\Psi = 0\,, \tag{33}$$

then eq. (32) can be reduced to the following form

$$\varrho^{-1}\bar{u}_{,\theta} = \varrho^{-1}(\varrho\bar{v})_{,\varrho} + \Psi\,. \tag{34}$$

Substituting this relation into eq. (31), the fundamental equation for $\bar{v}$ is

$$\bar{\Delta}(\varrho\bar{v}) = (1 + \nu)\,(1 - \nu)^{-1}\, \{\overline{T} - \Phi/(1 - 2\nu)\}_{,\theta} - 2\Psi\,. \tag{35}$$

These two equations (33) and (35) can be solved by means of successive integral transforms, and the solutions in transformed domain are shown as

$$\Psi^* = \sqrt{2/\pi} \int_0^\infty \sum_{n=1}^\infty C_n I_n(q\varrho) \sin n\theta \cos q\zeta \, dq , \tag{36}$$

$$\bar{v}^* = \sqrt{2/\pi} \int_0^\infty \sum_{n=1}^\infty \{D_n \varrho^{-1} I_n(q\varrho) - C_n q^{-1} I_{n+1}(q\varrho) - (1+\nu)(1-\nu)^{-1} [B_n(\mu^2 - q^2)^{-1} \varrho^{-1} n I_n(\mu\varrho) - \\ - \{1 + 2(1-\nu)\delta\} \{2(1-2\nu)(1+\delta)\}^{-1} A_n q^{-1} n I_{n+1}(q\varrho)]\} \sin n\theta \cos q\zeta \, dq , \tag{37}$$

in which C_n and D_n are also the unknown constants. Then $\bar{u}^*$ can be calculated from eq. (30) by using eq. (37), and shown as

$$\bar{u}^* = \sqrt{2/\pi} \int_0^\infty \left\{ \sum_{n=1}^\infty [-D_n\{\varrho^{-1} I_n(q\varrho) + q n^{-1} I_{n+1}(q\varrho)\}] \cos n\theta + \right. \\ + \sum_{n=0}^\infty \{-C_n q^{-1} I_{n+1}(q\varrho) + (1+\nu)(1-\nu)^{-1} [B_n(\mu^2 - q^2)^{-1} \{\varrho^{-1} n I_n(\mu\varrho) + \mu I_{n+1}(\mu\varrho)\} - \\ \left. - \{1 + 2(1-\nu)\delta\} \{2(1-2\nu)(1+\delta)\}^{-1} A_n\{\varrho I_n(q\varrho) - q^{-1} n I_{n+1}(q\varrho)\}]\} \cos n\theta \right\} \cos q\zeta \, dq . \tag{38}$$

Finally, the remaining component $\bar{w}^*$ can be calculated from eq. (19) and has the following form

$$\bar{w}^* = \sqrt{2/\pi} \int_0^\infty \left\{ \sum_{n=1}^\infty D_n q n^{-1} I_n(q\varrho) \cos n\theta + \sum_{n=0}^\infty \{C_n q^{-1} I_n(q\varrho) + (1+\nu)(1-\nu)^{-1} \times \right. \\ \times \{-B_n q(\mu^2 - q^2)^{-1} I_n(\mu\varrho) + A_n[2(1-\nu)(1-2\nu)^{-1} q^{-1} I_n(q\varrho) + \{1 + 2(1-\nu)\delta\} \times \\ \left. \times \{2(1-2\nu)(1+\delta)\}^{-1} \varrho I_{n+1}(q\varrho)]\}\} \cos n\theta \right\} \sin q\zeta \, dq . \tag{39}$$

Thus, all the displacements are obtained, then the stress components in the transformed domain are determined from the stress-strain relation (11). However, these precise descriptions cannot be given on account of the limited space.

In order to determine the integral constants A_n, B_n, C_n and D_n, we assume that the practical temperature distribution of the surrounding medium is given by the following manner. The temperature function $f(\theta)$ in tangential direction can be developed into the Fourier series form

$$f(\theta) = \sum_{n=0}^\infty a_n \cos n\theta . \tag{40}$$

Now, we assume that $f(\theta)$ is given by the step function as shown in Fig. 1, and the temperature function $g(\zeta)$ in lateral side is also represented by step function

$$f(\theta) = H(\theta_0 - |\theta|) , \qquad g(\zeta) = H(\zeta_0 - |\zeta|) , \tag{41}$$

in which $H(\cdot)$ denotes Heaviside's step function.

From the above considerations, all the integral constants, evaluated in the transformed domain, are uniquely determined for each value of $n (n \geqq 0)$ from the mechanical and thermal boundary conditions (22) and (23).

The required solutions of temperature and stress distributions can be obtained by means of inverse Laplace transform. We may perform the inverse transform by using the residue theorem, and find the temperature distribution and the associated normal stress distributions as follows

$$\bar{T} = 2H_a/\pi \cdot \int_0^\infty \sum_{n=0}^\infty a_n[I_n(q\varrho)/\{(n + H_a) I_n(q) + q I_{n+1}(q)\} - \sum_{i=1}^\infty 2\omega_{ni} G_n^1(\omega_{ni})/\{(\omega_{ni}^2 + q^2) H_n(\omega_{ni})\} \times \\ \times \exp\{-(\omega_{ni}^2 + q^2)\tau/(1+\delta)\}] \cos n\theta q^{-1} \sin q\zeta_0 \cos q\zeta \, dq ,$$

$$\bar{\sigma}_{rr} = 2H_a/\pi \cdot \int_0^\infty \sum_{n=0}^\infty a_n[F_n^2(q)/D_n(q) - (1-\nu)^{-1} \sum_{i=1}^\infty 2\omega_{ni} G_n^2(\omega_{ni})/\{(\omega_{ni}^2 + q^2) H_n(\omega_{ni})\} \times \\ \times \exp\{-(\omega_{ni}^2 + q^2)\tau/(1+\delta)\}] \cos n\theta q^{-1} \sin q\zeta_0 \cos q\zeta \, dq ,$$

$$\bar{\sigma}_{\theta\theta} = 2H_a/\pi \cdot \int_0^\infty \sum_{n=0}^\infty a_n[F_n^3(q)/D_n(q) - (1-\nu)^{-1} \sum_{i=1}^\infty 2\omega_{ni} G_n^3(\omega_{ni})/\{(\omega_{ni}^2 + q^2) H_n(\omega_{ni})\} \times \\ \times \exp\{-(\omega_{ni}^2 + q^2)\tau/(1+\delta)\}] \cos n\theta q^{-1} \sin q\zeta_0 \cos q\zeta \, dq , \tag{42}$$

$$\bar{\sigma}_{zz} = 2H_a/\pi \int_0^\infty \sum_{n=0}^\infty a_n[F_n^4(q)/D_n(q) - (1-\nu)^{-1} \sum_{i=1}^\infty 2\omega_{ni} G_n^4(\omega_{ni})/\{(\omega_{ni}^2 + q^2) H_n(\omega_{ni})\} \times \\ \times \exp\{-(\omega_{ni}^2 + q^2)\tau/(1+\delta)\}] \cos n\theta q^{-1} \sin q\zeta_0 \cos q\zeta \, dq ,$$

in which

$$F_n^2(q) = \varrho^{-2} n(n-1)\, I_n(q\varrho)\, C_1(q) + I_n(q\varrho)\, C_2(q) + q\varrho I_{n+1}(q\varrho)\, C_3(q) + (q\varrho)^{-1} I_{n+1}(q\varrho)\, C_4(q),$$

$$F_n^3(q) = -\varrho^{-2} n(n-1)\, I_n(q\varrho)\, C_1(q) - I_n(q\varrho)\, C_5(q) - (q\varrho)^{-1} I_{n+1}(q\varrho)\, C_4(q),$$

$$F_n^4(q) = I_n(q\varrho)\, C_6(q) - q\varrho I_{n+1}(q\varrho)\, C_3(q),$$

$$D_n(q) = C_7(q)\, \{C_8(q)\, C_9(q) + 2(1-\nu)\, C_3(q)\},$$

$$G_n^1(\omega_{ni}) = -(\omega_{ni}^2 + q^2)\, J_n(\omega_{ni}\varrho)\, \{K_1 C_8(q)\, C_9(q) + K_2 C_3(q)\} + K_3 q I_n(q\varrho)\, E_1(\omega_{ni})\, C_9(q),$$

$$\begin{aligned} G_n^2(\omega_{ni}) = {} & K_1 C_9(q)\, [[\{n(n-1)/\varrho^2 + q^2\}\, J_n(\omega_{ni}\varrho) + \omega_{ni}\varrho^{-1} J_{n+1}(\omega_{ni}\varrho)]\, C_8(q) - \\ & - [\{n(n-1)/\varrho^2 + q^2\}\, I_n(q\varrho) - q\varrho^{-1} I_{n+1}(q\varrho)]\, E_2(\omega_{ni}) - \\ & - q[I_n(q\varrho) - \{n(n+1)/(q\varrho) + q\varrho\}\, I_{n+1}(q\varrho)]\, E_1(\omega_{ni})] + \\ & + K_2[-[\{n(n-1)/\varrho^2 + q^2\}\, I_n(q\varrho) - q\varrho^{-1} I_{n+1}(q\varrho)]\, [J_n(\omega_{ni})\, C_{10}(q) - \omega_{ni} J_{n+1}(\omega_{ni})\, C_{11}(q)] - \\ & - 2qn\{I_n(q\varrho) - (n+1)/(q\varrho)\, I_{n+1}(q\varrho)\}\, E_1(\omega_{ni})\, C_{12}(q) + \\ & + [\{n(n-1)/\varrho^2 + q^2\}\, J_n(\omega_{ni}\varrho) + \omega_{ni}\varrho^{-1} J_{n+1}(\omega_{ni}\varrho)]\, C_3(q) + q I_n(q\varrho)\, E_1(\omega_{ni})\, C_9(q)]\,, \end{aligned}$$

$$\begin{aligned} G_n^3(\omega_{ni}) = {} & K_1 C_9(q)\, [\{n(n-1)\, \varrho^{-2} I_n(q\varrho) - q\varrho^{-1} I_{n+1}(q\varrho)\}\, E_2(\omega_{ni}) - [\{n(n-1)/\varrho^2 - \omega_{ni}^2 - q^2\}\, J_n(\omega_{ni}\varrho) + \\ & + \omega_{ni}\varrho^{-1} J_{n+1}(\omega_{ni}\varrho)]\, C_8(q) - q\{I_n(q\varrho) + n(n+1)\, (q\varrho)^{-1} I_{n+1}(q\varrho)\}\, E_1(\omega_{ni})] + \\ & + K_2[\{n(n-1)\, \varrho^{-2} I_n(q\varrho) - q\varrho^{-1} I_{n+1}(q\varrho)\}\, [J_n(\omega_{ni})\, C_{10}(q) - \omega_{ni} J_{n+1}(\omega_{ni})\, C_{11}(q)] - \\ & - 2n(n+1)\, \varrho^{-1} I_{n+1}(q\varrho)\, E_1(\omega_{ni})\, C_{12}(q) - [\{n(n-1)/\varrho^2 - \omega_{ni}^2 - q^2\}\, J_n(\omega_{ni}\varrho) + \\ & + \omega_{ni}\varrho^{-1} J_{n+1}(\omega_{ni}\varrho)]\, C_3(q) + q I_n(q\varrho)\, E_1(\omega_{ni})\, C_9(q)]\,, \end{aligned}$$

$$\begin{aligned} G_n^4(\omega_{ni}) = {} & K_1 C_9(q)\, [q^2 I_n(q\varrho)\, E_2(\omega_{ni}) + \omega_{ni}^2 J_n(\omega_{ni}\varrho)\, C_8(q) - q\{2 I_n(q\varrho) + q\varrho I_{n+1}(q\varrho)\}\, E_1(\omega_{ni})] + \\ & + K_2[q^2 I_n(q\varrho)\, \{J_n(\omega_{ni})\, C_{10}(q) - \omega_{ni} J_{n+1}(\omega_{ni})\, C_{11}(q)\} + 2qn I_n(q\varrho)\, E_1(\omega_{ni})\, C_{12}(q) + \\ & + \omega_{ni}^2 J_n(\omega_{ni}\varrho)\, C_3(q) - q I_n(q\varrho)\, E_1(\omega_{ni})\, C_9(q)]\,, \end{aligned}$$

$$\begin{aligned} H_n(\omega_{ni}) = {} & K_3 q C_7(q)\, C_9(q)\, [n\omega_{ni}^{-1}\{-q J_n(\omega_{ni})\, I_{n+1}(q) + \omega_{ni} J_{n+1}(\omega_{ni})\, I_n(q)\} + \\ & + q I_{n+1}(q)\, J_{n+1}(\omega_{ni}) - \omega_{ni} I_n'(q)\, J_n(\omega_{ni})] + [K_1 C_8(q)\, C_9(q) + K_2 C_3(q)] \times \\ & \times [2\omega_{ni}\{(n + H_a)\, J_n(\omega_{ni}) - \omega_{ni} J_{n+1}(\omega_{ni})\} - (\omega_{ni}^2 + q^2)[(n + H_a)\, \{-n\omega_{ni}^{-1} J_n(\omega_{ni}) + \\ & + J_{n+1}(\omega_{ni})\} - n J_{n+1}(\omega_{ni}) + \omega_{ni} J_n(\omega_{ni})]]\,, \end{aligned}$$

$$\begin{aligned} C_1(q) = {} & \{q^2 - n(n-1)\}\, q I_n^3(q) - \{(4n+1)\, q^2 - 2n(n^2-1)\}\, I_n^2(q)\, I_{n+1}(q) - \\ & - \{q^2 - (n+1)\, (5n-2)\}\, q I_n(q)\, I_{n+1}^2(q) + 2(n+1)\, q^2 I_{n+1}^3(q)\,, \end{aligned}$$

$$\begin{aligned} C_2(q) = {} & n(n-1)\, \{n(n-1) - q^2\}\, q I_n^3(q) + \{q^4 + 2n(n-1)\, (2n+1)\, q^2 - 2n^2(n^2-1) \times \\ & \times (n-2)\}\, I_n^2(q)\, I_{n+1}(q) + \{(n^2 - n - 1)\, q^2 - n(n^2-1)\, (5n-6)\}\, q I_n(q)\, I_{n+1}^2(q) - \\ & - 2(n+1)\, (n-1)^2\, q^2 I_{n+1}^3(q)\,, \end{aligned}$$

$$\begin{aligned} C_3(q) = {} & 2n^2(n-1)\, q I_n^3(q) - 2n(n-1)\, \{2n(n+1) - q^2\}\, I_n^2(q)\, I_{n+1}(q) - \{6n(n^2-1) + \\ & + q^2\}\, q I_n(q)\, I_{n+1}^2(q) - 2(n^2-1)\, q^2 I_{n+1}^3(q)\,, \end{aligned}$$

$$\begin{aligned} C_4(q) = {} & -\{q^4 + n(n-1)\, (2n+1)\, q^2 + 2n^2(n^2-1)\}\, q I_n^3(q) + \{(-2n^2 + 2n + 1)\, q^4 + 4n \times \\ & \times (n+1)\, (n-1)^2\, q^2\}\, I_n^2(q)\, I_{n+1}(q) + \{q^4 + 2(n^2-1)\, (3n-1)\, q^2\}\, q I_n(q)\, I_{n+1}^2(q) + \\ & + 2(n^2-1)\, q^4 I_{n+1}^3(q)\,, \end{aligned}$$

$$\begin{aligned} C_5(q) = {} & \{n^2(n-1)\, (n+3) + 2n(n-1)\, q^2 + q^4\}\, q I_n^3(q) - \{2n^2(n^2-1)\, (n+2) + 2n \times \\ & \times (n-1)\, (2n-1)\, q^2 + 2(n+1)\, q^4\}\, I_n^2(q)\, I_{n+1}(q) - \{n(n^2-1)\, (n+6) + \\ & + (2n^2 - 6n + 1)\, q^2 + q^4\}\, q I_n(q)\, I_{n+1}^2(q) + \{2q^2 - 2(n^2-1)\}\, q^2 I_{n+1}^3(q)\,, \end{aligned}$$

$$\begin{aligned} C_6(q) = {} & \{n(n-1)\, q^2 - n^2(n-1)\, (n+5)\}\, q I_n^3(q) + \{2n^2(n^2-1)\, (n+4) - 4n(n-1) \times \\ & \times (n+2)\, q^2 - q^4\}\, I_n^2(q)\, I_{n+1}(q) + \{n(n^2-1)\, (5n+12) + (-n^2 + n + 4)\, q^2\}\, q I_n(q)\, I_{n+1}^2(q) + \\ & + 2(n^2-1)\, (n+2)\, q^2 I_{n+1}^3(q)\,, \end{aligned}$$

$$C_7(q) = (n + H_a)\, I_n(q) + q I_{n+1}(q)\,,$$

$$C_8(q) = q I_n^2(q) - 2n I_n(q)\, I_{n+1}(q) - q I_{n+1}^2(q)\,,$$

$$C_9(q) = \{n(n+1) + q^2\}\, \{n(n-1) + q^2\}\, I_n(q) - 2q^2\{n I_n(q) + q I_{n+1}(q)\}\,,$$

$$\begin{aligned} C_{10}(q) = {} & 2n^2(n-1)\, q I_n^2(q) + \{q^4 + n(n-1)\, q^2 - 4n^2(n^2-1)\}\, I_n(q)\, I_{n+1}(q) - \\ & - 2(n+1)\, \{2n(n-1) + q^2\}\, q I_{n+1}^2(q)\,, \end{aligned}$$

$$C_{11}(q) = \{n(n-1) - q^2\}\, q I_n^2(q) + \{(2n+1)\, q^2 - 2n(n^2-1)\}\, I_n(q)\, I_{n+1}(q) - 2(n^2-1)\, q I_{n+1}^2(q)\,,$$

$$C_{12}(q) = (n-1)\, \{n(n+1) + q^2\}\, I_n(q) + \{(n^2-1) + q^2\}\, q I_{n+1}(q)\,,$$

$$E_1(\omega_{ni}) = qJ_n(\omega_{ni})\, I_{n+1}(q) + \omega_{ni}J_{n+1}(\omega_{ni})\, I_n(q)\,,$$

$$E_2(\omega_{ni}) = qJ_n(\omega_{ni})\, I_n(q) - 2nJ_n(\omega_{ni})\, I_{n+1}(q) + \omega_{ni}J_{n+1}(\omega_{ni})\, I_{n+1}(q)\,,$$

$$K_1 = \{1 + 2(1-\nu)\,\delta\}/\{2(1-2\nu)\,(1+\delta)\}\,, \qquad K_2 = (1-\nu)/(1-2\nu)\,, \qquad K_3 = \delta/(1+\delta)\,. \tag{43}$$

In the above expressions, the eigenvalues ω_{ni} should be determined as the i-th positive root of the following transcendental equation for each value of n ($n \geqq 0$)

$$K_3 q C_7(q)\, E_1(\omega)\, C_9(q) - (\omega^2 + q^2)\,\{(n + H_a)\, J_n(\omega) - \omega J_{n+1}(\omega)\}\,\{K_1 C_8(q)\, C_9(q) + {} + K_2 C_3(q)\} = 0\,. \tag{44}$$

As the quantity q is integral parameter, we have evaluated eq. (44) in the following manner, and determined the values ω_{ni} for the arbitrary values of q. For the numerical calculations of temperature and stress distributions given by eq. (42), it should be accomplished with aid of the method of numerical integration technique because these expressions are too complicated. Then, if we apply one of the closed type formula such as trapezoidal rule to carry out the numerical integration, we may observe that the integral parameter q may be regarded as a known value. Therefore, we have taken into account that the integral parameter q changes from 0 to 40 with its interval (mesh side) as 0.05, and calculated the eigenvalue ω_{ni} for each values of q. Thus, we have performed the numerical integration for eq. (42).

3. Numerical results and conclusion

To illustrate the foregoing analysis, numerical results are presented for the following values of material constants:

$$\nu = 0.3\,, \quad H_a = 10\,, \quad \zeta_0 = 1.0\,, \quad \theta_0 = \pi/12\,, \quad \delta = 0\,,\ 0.1 \tag{45}$$

The coupling parameter δ is taken into account for two cases. One of which is $\delta = 0$, which corresponds to the uncoupled thermoelastic problem, and another one is taken to be $\delta = 0.1$ in order to examine the effects of the thermocoupling term on the temperature and the stress distributions. This value of $\delta = 0.1$ is enough realistic and representative of commercial alloys. For instance, typical values of the coupling parameter δ under the condition of room temperature (20 °C) are

Aluminium alloy $= 0.03$, *Lead* $= 0.08$

Numerical results for the temperature and the stress distribution will be shown in figures. In all the figures subsequently shown, the dotted lines express the results for the coupled case of $\delta = 0.1$ and the full lines mean the uncoupled results of $\delta = 0$. And these figures are plotted for $\theta = 0$ with respect to tangential direction, because the temperature gradient becomes the maximum value in this cross section.

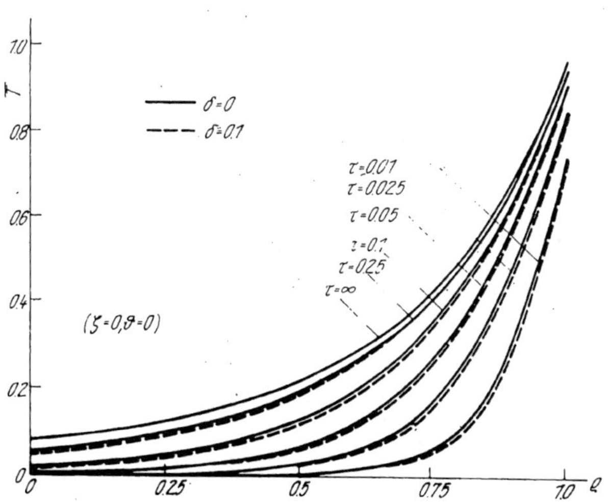

Fig. 2. Temperature variation in the cross sectional direction on the middle cross section ($\zeta = 0$, $\theta = 0$)

Fig. 2 shows temperature variation along the radial direction on the middle cross section ($\zeta = 0$) at various dimensionless times, and the axial variation of the temperature for several dimensionless radii ϱ are illustrated in Fig. 3. It can be seen from these figures that there is a considerable difference between the coupled and uncoupled results. In precisely, for the first short time period of heating, the difference is still pretty, however, in the next time interval of heating, this discrepancy becomes to be remarkable gradually. It seems from Fig. 2 that the thermocoupling term affects on distributions near the axis rather than near the surface with respect to radial variation as the time passes from initial state, and from Fig. 3 the thermocoupling effect with respect to the lateral variation is remarkable within a region received the heat supply.

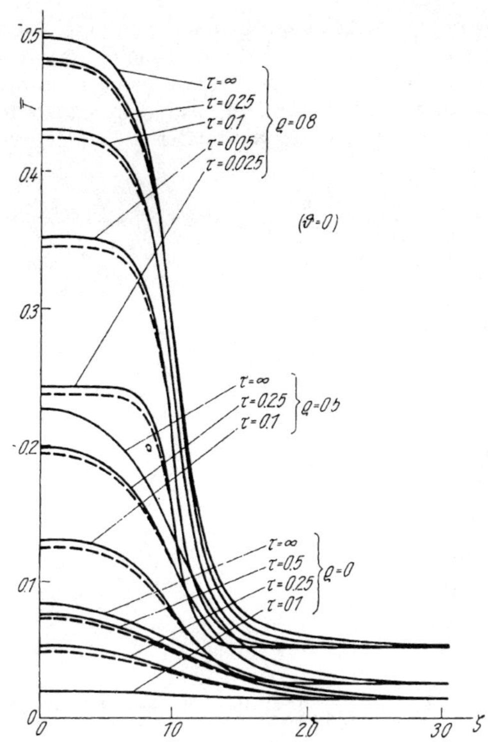

Fig. 3. Temperature variation in the lateral direction ($\theta = 0$, $\varrho = 0, 0.5, 0.8$)

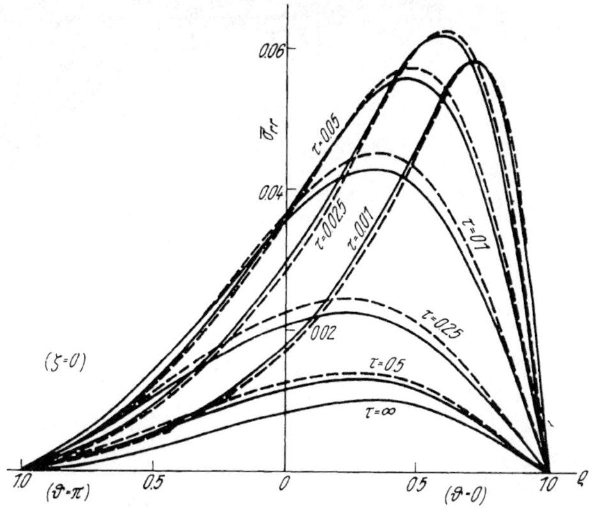

Fig. 4. Radial stress variation in the cross sectional direction on the middle cross section ($\zeta = 0$, $\theta = 0, \pi$)

Fig. 4 shows a variation of radial stress $\bar{\sigma}_{rr}$ on the middle cross section ($\zeta = 0$) along the axis $\theta = 0$ and $\theta = \pi$. This radial stress distribution is considerably small compared with the other normal stress $\bar{\sigma}_{\theta\theta}$ or $\bar{\sigma}_{zz}$, because the temperature distribution of surrounding medium is considered to be very local in the present numerical example, however, the thermocoupling effect is explicitly appeared for the radial stress distribution.

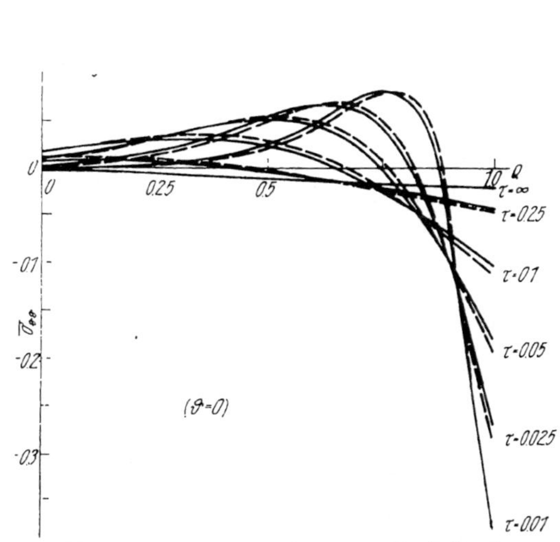

Fig 5. Hoop stress variation in the cross sectional direction on the middle cross section ($\zeta = 0$, $\theta = 0$)

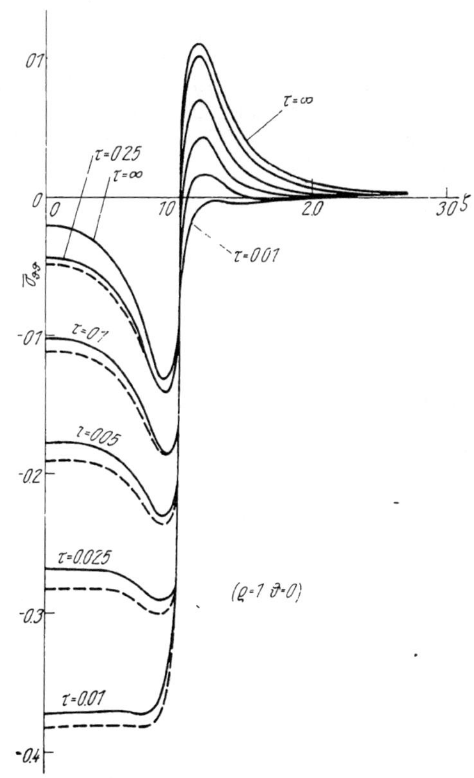

Fig. 6. Hoop stress variation along the lateral surface ($\theta = 0$, $\varrho = 1$)

Figs. 5 and 6 show the variations of hoop stress $\bar{\sigma}_{\theta\theta}$, the radial variation on the middle cross section is illustrated in Fig. 5, and the axial variations along the lateral surface is shown in Fig. 6. From these figures, it can be seen that the maximum compressive stress of $\bar{\sigma}_{\theta\theta}$ at the surface of the middle cross section is considerably increased when the coupling effect is taken into account, but, the maximum tensile stress appeared in the adjacent region near the surface is almost unvaried in accordance with the change of coupling parameter.

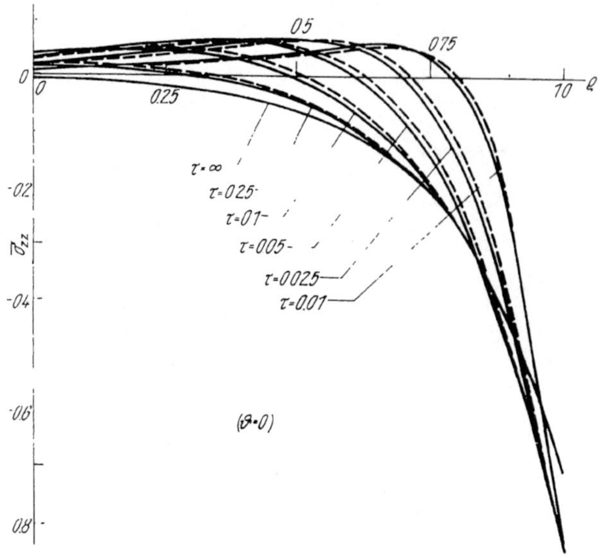

Fig. 7. Axial stress variation in the cross sectional direction on the middle cross section ($\zeta = 0$, $\theta = 0$)

Finally, the radial variation of axial stress $\bar{\sigma}_{zz}$ in the middle cross section is shown in Fig. 7. We may observe from this figure that the maximum compressive stress appeared on the surface is hardly varied when the coupling effect is taken into account.

In conclusion, we may summarize the effect of the thermocoupling term from our numerical calculation as follows. When the thermocoupling effect is taken into account, the temperature distribution is decreased in comparison with the uncoupled result because of the contribution of the heat energy to the dissipative strain energy. However, as for the stress distributions, it becomes to be complicated, the hoop stress distribution is increased considerably for its maximum compressive stress distribution, on the other side, the axial stress distribution becomes to give still large values, however, the maximum stress is almost unvaried. And the thermocoupling effect in axial direction is bounded within a region of heat supply by the surrounding medium.

In the present paper, we have discussed about the three-dimensional coupled thermoelastic problem, and have proposed a new technique which is possible to analyze both the temperature field and the associated stress fields simultaneously, and have carried out the numerical calculation for the solid cylinder under three-dimensional temperature field. However, our new analytical technique proposed in the present paper is available for other orthogonal coordinate systems, then we may expect that the present method can be used extensively to the other coupled problems.

Reference

1 Takeuti, Y.; Furukawa, T., Some consideration on thermal shock problem in a plate, Journal of applied Mechanics, 48-2 (1981), pp. 113–118.

Eingereicht am 5. 1. 1981

Anschrift: Prof. Y. Takeuti, Department of Mechanical Engineering, University of Osaka Prefecture, Mozu-umemachi, Sakai 591, Osaka, Japan

ZAMM 62, 229 – 240 (1982)

J. J. Sławianowski

The Mechanics of the Homogeneously-Deformable Body. Dynamical Models with High Symmetries

We investigate here some high-symmetric isotropic models in the theory of nonlinear vibrations of systems with degrees of freedom governed by a linear group (affinely-rigid bodies undergoing gyroscopic rotations and homogeneous deformations). The methods developed here can be applied both to microscopic (molecular, or even nuclear) and macroscopic elastic and visco-elastic problems. In particular, our models can be used in the theory of generalized, micromorphic continua consisting of infinitesimal, homogeneously-deformable grains.

Wir untersuchen hier einige hochsymmetrische isotrope Modelle in der Theorie nichtlinearer Schwingungen von Systemen. deren Freiheitsgrade durch eine lineare Gruppe bestimmt werden (affin-starre Körper, die gyroskopischen Drehungen und homogenen Deformationen unterworfen sind). Die hier entwickelten Methoden können sowohl auf mikroskopische (molekulare oder sogar nukleare) als auch auf makroskopische elastische und visko-elastische Probleme angewandt werden. Insbesondere können unsere Modelle in der Theorie verallgemeinerter mikromorpher Kontinua verwendet werden, die aus infinitesimalen, homogen-deformierbaren Körnern bestehen.

Рассматриваются некоторые высокосимметричные модели в теории нелинейных колебаний систем, степени свободы которых определяются с помощью линейной группы (аффинно-твердые тела, на которые действуют гироскопические вращения и однородные деформации). Здесь развитые методы можно применять и на микроскопические (молекулярные и даже ядерные) и на макроскопические упругие и вязко-упругие задачи. В особенности наши модели можно применять в теории обобщенных микроморфических сред, которые состоят из бесконечно малых однородно-деформируемых тел.

1. Introduction

The general mechanical theory of systems endowed only with affine modes of deformations has been developed in our earlier papers [11], [12], [13], [15]. The exhaustive treatment of the geometrical background of the theory was given there, especially in [13], [15]. In the present paper we intend to investigate qualitatively the doubly-isotropic dynamical models, having in view applications in the theory of molecular and nuclear vibrations and in some problems of macroscopic elasticity (both classical and generalized, micromorphic). The model is applicable in all situations where the interaction radius and typical wave lengths are comparable with linear sizes of the body. It has been also successfully applied in some astrophysical problems [2], [3], [4], [5].

We are interested in qualitative properties of the particular dynamical model rather than in geometric foundations of the theory (analysed in [13], [15]). That is why we prefer to use here the matrix notation instead of abstract terms employed in [11], [13], [15].

Let us consider an n-dimensional body subject to constraints which confine its deformative behaviour to homogeneous strains. We do not take into account the motion of the centre of mass. Any particular choice of the Cartesian coordinate system in the physical space gives rise to some identification of the configuration space of the homogeneously-deformable body with $GL(n, R)$ — the group of non-singular real $n \times n$ matrices. To be general, we do not precise at this stage the value of n; of course, only the special cases $n = 1, 2, 3$ are physically interesting. Such a description of the configuration space is to be understood in such a way that any $\varphi \in GL(n, R)$ connects the Eulerian and Lagrangean Cartesian coordinates:

$$x^i = \sum_j \varphi^{ij} X^j \tag{1.1}$$

($X^j, x^j, j = 1, \dots, n$, are respectively Cartesian coordinates of the same physical particle in the reference and actual configurations).

$GL(n, R)$ is an open submanifold of $L(n, R)$ — the linear space of all real $n \times n$ matrices. Hence, all sophisticated techniques of the calculus on manifolds can be avoided — the tangent and cotangent bundles over $Q = GL(n, R)$ both become canonically identified with $GL(n, R) \times L(n, R)$.

The configuration space $GL(n, R)$ carries two natural transformation groups, namely the left and right actions of $GL(n, R)$ itself:

$$\varphi \mapsto A\varphi\,, \qquad \varphi \mapsto \varphi A \tag{1.2}$$

where $A \in GL(n,R)$. These groups commute with each other. Their intersection consists only of dilatations:

$$\varphi \mapsto \lambda\varphi\,, \qquad \lambda \in R^+\,.$$

The left hand side actions describe homogeneous deformations in the physical space and right actions describe transformations of the material itself.

Similarly, the orthogonal subgroup $SO(n, R) \subset GL(n, R)$ (the group of orthogonal matrices) acts through the above formulae as a group of physical (left) and material (right) rotations.

Green and Cauchy deformation tensors, G, C are given by the following formulas:

$$G = \varphi^T\varphi\,, \qquad C^{-1} = \varphi\varphi^T\,. \tag{1.3}$$

Obviously, $G^T = G$, $C^T = C$. The Green matrix G is non-sensitive to the left orthogonal mappings and similarly C is invariant under the material rigid rotations. Of course, the deformation invariants, i.e. eigenvalues of the symmetric matrix G (or, equivalently C^{-1}) are invariant under both space and material rigid rotations.

One uses often the following invariant quantities:

$$\varkappa_a = \mathrm{Tr}\,(G^a)\,, \qquad a = 1, \dots, n\,. \tag{1.4}$$

When $\lambda_1, \dots, \lambda_n$ are eigenvalues of G, then, obviously: $\varkappa_a = \sum_{i=1}^{n} \lambda_i^a$. Of course, both systems of invariants, $(\varkappa_1, \dots, \varkappa_n)$, $(\lambda_1, \dots, \lambda_n)$, are equivalent. One uses often the other system of invariants I_k consisting of coefficients of the eigen-equation for G:

$$\mathrm{Det}\,\|D - \lambda I\| = 0, \quad \text{i.e.:} \quad \sum_{k=0}^{n} (-1)^k\, I_{n-k} \lambda^k = 0\,; \tag{1.5}$$

I_{n-k} is a sum of products of $n - k$ different eigenvalues, in particular: $I_0 = 1$, $I_1 = \sum_i \lambda_i, \dots, I_n = \lambda_1 \dots \lambda_n$.

Motions are described by smooth curves in $GL(n, R)$, i.e. by time-dependent matrices $t \mapsto \varphi(t) \in GL(n, R)$. Generalized velocity: $\xi(t) = \frac{\mathrm{d}\varphi}{\mathrm{d}t} \in L(n, R)$ is a gradient of the Lagrangean velocity field V:

$$V^i(X) = \sum_j \frac{\mathrm{d}\varphi^{ij}}{\mathrm{d}t} X^j \tag{1.6}$$

(V attributes to the material point with reference coordinates X^i, its velocity at a given instant).

Eulerian velocity field is related to the Lie-algebraic structures of $GL(n, R)$. Namely, for arbitrary configuration $\varphi \in GL(n, R)$ and generalized velocity $\xi \in L(n, R)$ we define the non-holonomic velocity Ω called by Eringen gyration [6]:

$$\Omega = \xi\varphi^{-1}\,, \tag{1.7}$$

and its co-moving representation:

$$^c\Omega = \varphi^{-1}\xi = \varphi^{-1}\Omega\varphi\,. \tag{1.8}$$

This is equivalent to replacing generalized velocities by elements of the Lie algebra of $GL(n, R)$ which is canonically identified with a commutator matrix algebra $L(n, R)$. Such a description is common for mechanical systems with degrees of freedom governed by a symmetry Lie group [1].

In the special case of rigid motion, when $\varphi \in SO(n, R)$, the matrices Ω, $^c\Omega$ become skew-symmetric and their independent elements are just the components of the angular velocity (with respect to the space-fixed and body-fixed reference frames, respectively).

Gyration Ω is a gradient of the Eulerian velocity field v:

$$v^i(x) = \sum_j \Omega^{ij} x^j\,, \tag{1.9}$$

($v(x)$ is the velocity of the material point which occupies the space position with coordinates x^i).

The quantity Ω is invariant under right, material deformations $\varphi \mapsto \varphi A$ and transforms according to the adjoint rule:

$$\Omega \mapsto A\Omega A^{-1}\,, \tag{1.10}$$

under physical, i.e. left deformations: $\varphi \mapsto A\varphi$.

Contrary, $^c\Omega$ is invariant under space deformations and transforms according to the adjoint rule:

$$^c\Omega \mapsto A^{-1}\,{}^c\Omega A \tag{1.11}$$

under material affine rotations $\varphi \mapsto \varphi A$.

Internal inertial properties of the homogeneously-deformable body are described by the constant, symmetric matrix of the co-moving coefficients of inertia J. By an appropriate choice of Lagrangean coordinates we can always achieve the diagonal form of J. The matrix J is in the following way related to the Lagrangean mass distribution:

$$J^{ij} = \int X^i X^j \,\mathrm{d}\mu(X) \tag{1.12}$$

where the measure μ describes the mass distribution in the reference configuration.

The inertial object J is in the following way related to the quantity $\mathcal{J}$ which is usually used as the measure of rotational inertia of gyroscopic systems (moment of inertia):

$$\mathcal{J} = \mathrm{Tr}\,(J)\, I - J\,. \tag{1.13}$$

The kinetic energy:

$$T = \tfrac{1}{2} \int \sum_i V^i(X)\, V^i(X)\, \mathrm{d}\mu(X) \tag{1.14}$$

and the matrix of the first virial coefficients of the momentum:

$$K^{ij} = \int x^i(X)\, V^j(X)\, \mathrm{d}\mu(X) \tag{1.15}$$

may be rewritten in the following matrix form:

$$T = \frac{1}{2}\operatorname{Tr}\left(J\frac{\mathrm{d}\varphi^T}{\mathrm{d}t}\frac{\mathrm{d}\varphi}{\mathrm{d}t}\right), \qquad K = \varphi J\frac{\mathrm{d}\varphi^T}{\mathrm{d}t}. \tag{1.16}$$

The quantity K is called by HEHL hypermomentum [7], [8] and by RÖMER and WESTPFAHL [10] — affine spin. The latter name has been used independently in our papers [13], [15].

The matrix of internal angular momentum (spin) has a form:

$$S = K - K^T = \varphi J\frac{\mathrm{d}\varphi^T}{\mathrm{d}t} - \frac{\mathrm{d}\varphi}{\mathrm{d}t}J\varphi^T. \tag{1.17}$$

Let F be the Eulerian density of forces per unit mass. The matrix of the first virial coefficients of forces is given as follows:

$$N^{ij} = \int x^i(X)\, F^j\left(\varphi\cdot X, \frac{\mathrm{d}\varphi}{\mathrm{d}t}\cdot X, t\right)\mathrm{d}\mu(X) = \sum_k \varphi^{ik}\int X^k F^j\left(\varphi\cdot X, \frac{\mathrm{d}\varphi}{\mathrm{d}t}\cdot X, t\right)\mathrm{d}\mu(X). \tag{1.18}$$

The twice skew-symmetric part of N, $N - N^T$ is the usual momentum of forces.

After performing the integration over material variables X we obtain the expression of N as a function of generalized coordinates φ, velocities $\mathrm{d}\varphi/\mathrm{d}t$ and time t. In continuum theory with short-range contact interactions internal forces are derived from the Cauchy stress tensor and then:

$$N = -\int \sigma \tag{1.19}$$

where the integration is extended all over the whole physical space (as a matter of fact — over the support of σ). In the non-polar case both σ, N are symmetric.

When there is no dissipation and the equations of motion are derivable from the Lagrangean $L = T - V$ with velocity-independent potential energy V, then:

$$N^{ij} = -\sum_k \varphi^{ik}\frac{\partial V}{\partial \varphi^{jk}}. \tag{1.20}$$

In the papers [12], [15] we have shown that the D'ALEMBERT principle leads to the following equations of motion:

$$\varphi J\frac{\mathrm{d}^2\varphi^T}{\mathrm{d}t^2} = N \tag{1.21}$$

or, equivalently:

$$\frac{\mathrm{d}^2\varphi}{\mathrm{d}t^2}J\varphi^T = N^T. \tag{1.22}$$

Therefore, the skew-symmetric part of the equations of motion is equivalent to the balance law for spin S:

$$\frac{\mathrm{d}S}{\mathrm{d}t} = N - N^T \tag{1.23}$$

and becomes the spin conservation law when N is symmetric. Similarly, the complete system of equations of motion can be rewritten as the balance law for the virial coefficients K:

$$\frac{\mathrm{d}K}{\mathrm{d}t} = N + \frac{\mathrm{d}\varphi}{\mathrm{d}t}J\frac{\mathrm{d}\varphi^T}{\mathrm{d}t}. \tag{1.24}$$

However, even in the interaction-free case, when $N = 0$, the quantity K is non-conserved. The reason is that the affine symmetry of degrees of freedom is broken by the kinetic energy and restricted to the more narrow Euclidean subgroup $SO(n, R) \subset GL(n, R)$ [11], [12], [13], [15].

The balance form of equations of motion is especially useful when investigating the motion of an affinely-rigid body subject to some additional constraints. For example, the obvious group-theoretical analysis shows that the motion of an incompressible affinely-rigid body is described by the trace-less part of (1.21). The configuration space is to be identified then with the unimodular group $SL(n, R) \subset GL(n, R)$. Similarly, when the body undergoes the shape-preserving motions, its configuration space becomes $SO(n, R) \times R^+ = \{\lambda R : R \in SO(n, R), \lambda > 0\}$ and the complete system of equations of motion consists of the skew-symmetric part and the trace of (1.21). Finally, the dynamics of rigid, gyroscopic motion is described by the skew-symmetric part of (1.21) and the configuration space becomes then $SO(n, R)$.

2. General Description of Isotropic Models

When there are no external forces, equations of motion should be invariant under the group of rigid rotations of the physical space. This groups consists of mappings $\varphi \mapsto R\varphi$, where $R \in SO(n, R)$. Therefore, the invariance conditions take on the following form:

$$N\left(R\varphi, R\frac{\mathrm{d}\varphi}{\mathrm{d}t}\right) = RN\left(\varphi, \frac{\mathrm{d}\varphi}{\mathrm{d}t}\right)R^T \tag{2.1}$$

for arbitary $R \in SO(n, R)$. In the special case of velocity-independent forces this implies that the virial coefficients N have the form:

$$N(\varphi) = \varphi F(G) \varphi^T \tag{2.2}$$

where F is a matrix-valued function of the GREEN deformation matrix $G = \varphi^T \varphi$.

The equations of motion derivable from the Lagrangean $L = T - V$ with velocity-independent potential V (no magnetic forces) are invariant under physical rotations if and only if the potential is invariant (this follows from the invariance of the kinetic energy). Hence, the potential energy should be a function of the GREEN matrix, $V(G)$ and

$$N = -\varphi(D_G V) \varphi^T \tag{2.3}$$

where the symmetric matrix $D_G V$ is defined by the following formula:

$$\frac{\mathrm{d}}{\mathrm{d}t} V(G + tg)\bigg|_{t=0} = \mathrm{Tr}\,(g^T D_G V) \tag{2.4}$$

for an arbitrary matrix g. This agrees with the above formula for N.

Especially interesting is the special, high-symmetric case when the system is doubly isotropic, i.e. invariant under both physical and material rigid rotations:

$$\varphi \mapsto O\varphi R\,, \qquad O, R \in SO(n, R)\,. \tag{2.5}$$

Invariance of equations of motion under material rotations means that

$$RJR^T = J\,, \qquad N\left(\varphi R, \frac{\mathrm{d}\varphi}{\mathrm{d}t} R\right) = N\left(\varphi, \frac{\mathrm{d}\varphi}{\mathrm{d}t}\right), \tag{2.6}$$

for arbitrary $R \in SO(n, R)$. Therefore, J is proportional to the identity matrix:

$$J = \mu I \tag{2.7}$$

and all moments of inertia are equal.

In the special case of the velocity-independent internal interactions the second of the above invariance equations (2.6) means that N depends on φ through the CAUCHY matrix $C^{-1} = \varphi\varphi^T : N(C)$. The simultaneous space and material invariance conditions imply N to have the structure of a power series in C^{-1}, the coefficients being scalar functions of the basic invariants. The classical CAYLEY-HAMILTON theorem reduces this power series to a polynomial of degree n:

$$N = \sum_{a=1}^{n} B_a C^{-a} = \varphi \left[\sum_{a=1}^{n} B_a G^{a-1} \right] \varphi^T \tag{2.8}$$

where B_a are real-valued functions of invariants.

When equations of motion result from a Lagrangean $L = T - V$ with velocity-independent potential V, then the joint conditions of physical and material isotropy imply again that $J = \mu I$ and that V depends on invariants only. In the above formula for N one should substitute then

$$B_a = -2a \frac{\partial V}{\partial \varkappa_a}. \tag{2.9}$$

In practically useful dynamical models N should consist of two parts:

1) the reversible one, N_{rev} independent of velocities (and usually derivable from some potential energy V),
2) the dissipative part N_{dis}, depending on $\mathrm{d}\varphi/\mathrm{d}t$ and responsible for the entropy production.

We have just seen what is the general form of N_{rev} compatible with requirements of the double isotropy. What concerns the dissipative part, N_{dis}, the simplest and at the same time the most useful model is that of internal viscous friction linear in velocities.

The phenomenological viscous stress tensor is given by:

$$\sigma_{\mathrm{vis}} = 2\eta d + \left(\zeta - \frac{2\eta}{n}\right) \mathrm{Tr}(d)\, I \tag{2.10}$$

where d is the deformation rate, i.e. the symmetrized gradient of the Eulerian velocity field. In the special case of a homogeneously deformable body:

$$d = \tfrac{1}{2}(\Omega + \Omega^T) \tag{2.11}$$

and hence

$$\sigma_{\mathrm{vis}} = \eta(\Omega + \Omega^T) + \left(\zeta - \frac{2\eta}{n}\right) \mathrm{Tr}\,(\Omega)\, I \tag{2.12}$$

where η, ζ are viscous coefficients. Hence, according to the formula $N = -\int \sigma$, we have:

$$N_{\mathrm{dis}} = -V_0 \,\mathrm{Det}\,\varphi \left[\eta(\Omega + \Omega^T) + \left(\zeta - \frac{2\eta}{n}\right) \mathrm{Tr}\,(\Omega)\, I\right] \tag{2.13}$$

where V_0 is the undeformed volume of the body.

It is obvious that such a form of $N_{\rm dis}$ is compatible with our requirements of physical and material isotropy:

$$N_{\rm dis}\left(\varphi R, \frac{d\varphi}{dt} R\right) = N_{\rm dis}\left(\varphi, \frac{d\varphi}{dt}\right) \tag{2.14}$$

because Ω is invariant under all material affine rotations (in particular under rigid, Euclidean rotations),

$$RN_{\rm dis}\left(\varphi, \frac{d\varphi}{dt}\right) R^T = N_{\rm dis}\left(R\varphi, R\frac{d\varphi}{dt}\right), \tag{2.15}$$

what follows from the transformation properties of Ω under physical affine rotations.

In a linear infinitesimal theory we put $\varphi = I + \alpha$, where the matrix $\alpha \in L(n, R)$ is assumed to be small. It is convenient to split α into symmetric and skew-symmetric parts:

$$\alpha = e + r\,, \qquad e^T = e\,, \qquad r^T = -r\,. \tag{2.16}$$

Then $G \approx (I + 2e)$ and e is the EULER infinitesimal deformation tensor. The dynamics of the isotropic continuum is based on the well known Hooke'an form of the stress tensor:

$$\sigma_{\rm el} = 2\mu e + \lambda \operatorname{Tr}(e)\, I\,. \tag{2.17}$$

Hence

$$N_{\rm el} = -2V_0\mu e - V_0\lambda \operatorname{Tr}(e)\, I = -2V_0\mu e - V_0\left(K - \frac{2\mu}{n}\right)\operatorname{Tr}(e)\, I \tag{2.18}$$

and

$$N_{\rm dis} = -2V_0\eta\frac{de}{dt} - V_0\left(\zeta - \frac{2\eta}{n}\right)\operatorname{Tr}\left(\frac{de}{dt}\right) I \tag{2.19}$$

provided we use (2.12) as a viscous stress tensor.

Let J be an arbitrary inertial matrix, not necessarily isotropic (the body is isotropic from the microscopic point of view, but we admit it to be anisotropic on the macroscopic level of global inertial properties). The equations of motion become then:

$$\frac{d^2e}{dt^2} = -V_0\mu(J^{-1}e + eJ^{-1}) - V_0\eta\left(J^{-1}\frac{de}{dt} + \frac{de}{dt}J^{-1}\right) - $$
$$- V_0\left(K - \frac{2\mu}{n}\right)\operatorname{Tr}(e)\, J^{-1} - V_0\left(\zeta - \frac{2\eta}{n}\right)\operatorname{Tr}\left(\frac{de}{dt}\right)J^{-1}\,, \tag{2.20}$$

$$\frac{d^2r}{dt^2} = V_0\mu(J^{-1}e - eJ^{-1}) + V_0\eta\left(J^{-1}\frac{de}{dt} - \frac{de}{dt}J^{-1}\right). \tag{2.21}$$

The nice feature of the above system is that equations of motion for the deformation e form the closed, autonomous system independent on rotations r. Contrary, the right hand side of (2.21) depends on e only and r appears there only through the acceleration term d^2r/dt^2. Hence, the deformative vibrations can be found from the first subsystem alone. To find rotational vibrations one should substitute the solution for $e(t)$ into the right hand side of (2.21) and integrate it twice. It is necessary to admit only such integration constants that the corresponding solution $r(t)$ is compatible with the model of small vibrations.

When the body is not only microscopically, but also macroscopically isotropic, then

$$J = bI \tag{2.22}$$

and our equations of motion reduce to

$$b\frac{d^2e}{dt^2} = -2V_0\mu e - 2V_0\eta\frac{de}{dt} - V_0\left(K - \frac{2\mu}{n}\right)\operatorname{Tr}(e)\, I - V_0\left(\zeta - \frac{2\eta}{n}\right)\operatorname{Tr}\left(\frac{de}{dt}\right) I\,, \qquad \frac{d^2r}{dt^2} = 0\,. \tag{2.23}$$

The only solutions of the second equation compatible with the initial infinitesimal assumptions have the form $r = \text{const}$. Deformative equations can be split into shear and volume parts:

$$b\frac{d^2}{dt^2}\left(e - \frac{1}{n}\operatorname{Tr}(e)\, I\right) = -2V_0\mu\left(e - \frac{1}{n}\operatorname{Tr}(e)\, I\right) - 2V_0\eta\frac{d}{dt}\left(e - \frac{1}{n}\operatorname{Tr}(e)\, I\right),$$
$$b\frac{d^2}{dt^2}\operatorname{Tr}(e) = -nV_0K\operatorname{Tr}(e) - nV_0\zeta\frac{d}{dt}\operatorname{Tr}(e)\,. \tag{2.24}$$

These subsystems are independent and are solved separately. The motion is highly degenerate — we have six deformative degrees of freedom but only two distinct frequencies and damping constants.

3. The Adapted Description of Degrees of Freedom. Structure of Equations of Motion

The most natural coordinates on $GL(n, R)$, namely the matrix elements of matrices φ, are inconvenient when studying doubly-isotropic dynamical models. The most adequate description of degrees of freedom is that based on the following decomposition of non-singular matrices:

$$\varphi = ODR^T \tag{3.1}$$

where $O,\ R \in SO(n, R)$ are special orthogonal matrices ($O^T O = R^T R = I$, $\det O = \det R = 1$) and D is diagonal and positive. This representation is unique up to the permutation of diagonal elements of D accompanied by the simultaneous multiplying of O and R (on the right) by the appropriate special orthogonal matrix. This implies that the potential energy of doubly isotropic models depends only on D and is invariant with respect to the permutations of its non-vanishing matrix elements.

Deformation matrices become now:

$$G = RD^2R^T\,, \qquad C^{-1} = OD^2O^T\,. \tag{3.2}$$

Therefore, the squared elements of D are equal to the eigenvalues of the Green deformation matrix; D is one of the equivalent systems of basic invariants. The scalar state of deformation is described by D, and orthogonal matrices O, R describe the space and body orientation of the strain. Hence, we are dealing with three kinds of degrees of freedom:

— pure scalar deformations D
— Cauchy gyroscope — the triad of principal axes of C^{-1} described by O
— Green gyroscope — the triad of principal axes of G, described by R.

The motion is represented by the time-dependence of these objects, i.e. by smooth curves: $t \mapsto D(t)$, $t \mapsto R(t)$, $t \mapsto O(t)$.

Instead of using generalized velocities of gyroscopic motions we will employ non-holonomic objects, namely, the angular velocities:

$$_{l}X = \frac{\mathrm{d}O}{\mathrm{d}t}\,O^T\,, \qquad _{l}\Theta = \frac{\mathrm{d}R}{\mathrm{d}t}\,R^T\,. \tag{3.3}$$

These matrices are skew-symmetric:

$$_{l}X^T = -_{l}X\,, \qquad _{l}\Theta^T = -_{l}\Theta \tag{3.4}$$

and their independent elements are the components of angular velocities with respect to the space-fixed, laboratory frame. Dynamical equations are based on the co-moving description:

$$X = O^T\,\frac{\mathrm{d}O}{\mathrm{d}t}\,, \qquad \Theta = R^T\,\frac{\mathrm{d}R}{\mathrm{d}t}\,. \tag{3.5}$$

Obviously:

$$X^T = -X\,, \qquad \Theta^T = -\Theta\,. \tag{3.6}$$

Independent elements of X are components of the angular velocity of the Cauchy gyroscope with respect to the C-co-moving frame built of the principal axes of C. Similarly, Θ is a system of co-moving components of the angular velocity of the Green gyroscope.

The theory of doubly-isotropic dynamical models leads to the closed system of equations for D and the co-moving angular velocities X, Θ. Configurations of G- and C-gyroscopes do not enter into these equations. In fact, the straightforward calculations show that the kinetic energy has a form

$$T = -\frac{\mu}{2}\,\mathrm{Tr}\,(D^2X^2) - \frac{\mu}{2}\,\mathrm{Tr}\,(D^2\Theta^2) + \mu\,\mathrm{Tr}\,(DXD\Theta) + \frac{\mu}{2}\,\mathrm{Tr}\left(\frac{\mathrm{d}D}{\mathrm{d}t}\right)^2 \tag{3.7}$$

where μ is the inertial parameter:

$$J = \mu I\,. \tag{3.8}$$

Similarly, our equations of motion can be rewritten as follows:

$$D^2X^2 - D^2\,\frac{\mathrm{d}X}{\mathrm{d}t} - 2D\,\frac{\mathrm{d}D}{\mathrm{d}t}\,X - 2D\Theta DX + D\Theta^2D + D\,\frac{\mathrm{d}\Theta}{\mathrm{d}t}\,D + 2D\Theta\,\frac{\mathrm{d}D}{\mathrm{d}t} + D\,\frac{\mathrm{d}^2D}{\mathrm{d}t^2}$$
$$= \frac{1}{\mu}\,B = \frac{1}{\mu}\,O^TNO\,. \tag{3.9}$$

The dynamical quantity $B = O^TNO$ on the right hand side of these equations contains the symbol O, but, as a matter of fact, the conditions of a double isotropy imply B to be independent on O and R.

For example, in the case of static (velocity-independent) interactions

$$N = \sum_{a=1}^{n} B_aC^{-a} = \varphi\left[\sum_{a=1}^{n} B_aG^{a-1}\right]\varphi^T = O\left[\sum_{a=1}^{n} B_aD^{2a}\right]O^T \tag{3.10}$$

where scalars B_a depend on D only. Hence, the quantity

$$B = O^TNO = \sum_{a=1}^{n} B_aD^{2a} \tag{3.11}$$

is a diagonal matrix independent on gyroscopic quantities O, R. Similarly, the dissipative term N_{dis} leads to the quantity

$$B_{\mathrm{dis}} = O^TN_{\mathrm{dis}}O = -V_0\eta\,\det D\left[2D^{-1}\,\frac{\mathrm{d}D}{\mathrm{d}t} - D\Theta D^{-1} + D^{-1}\Theta D\right] - V_0\left(\zeta - \frac{2\eta}{n}\right)\det D\,\mathrm{Tr}\left[D^{-1}\,\frac{\mathrm{d}D}{\mathrm{d}t}\right]I \tag{3.12}$$

where obviously $\det D = D_1 \dots D_n = I_n$. The quantity B_{dis} does depend on the angular velocity but nevertheless it is still independent on gyroscopic configurations O, R.

Thus, our task to solve the equations of motion

$$\varphi J \frac{d^2\varphi^T}{dt^2} = N \tag{3.13}$$

splits into two steps:
1. To solve the equations (3.9) with respect to D, X, Θ, i.e. to find the time dependence of parameters D, X, Θ.
2. To substitute the functions of time X, Θ found above into the equations

$$O^T \frac{dO}{dt} = X\,, \qquad R^T \frac{dR}{dt} = \Theta\,. \tag{3.14}$$

These are non-autonomous dynamical systems (first-order, ordinary differential equations) over the orthogonal group $SO(n, R)$. Solving them we find the time dependence of matrices O, R. The time evolution of D is known from the step 1, hence, in this way we achieve the final solution

$$\varphi(t) = O(t)\, D(t)\, R^T(t)\,. \tag{3.15}$$

In general, our system of nonlinear differential equations (3.9) is too complicated to be solved explicitly and one has to employ the qualitative methods of the theory of dynamical systems and some numerical techniques.

It happens often that we are interested not in the general solution of the problem but in some particular solution of special kind. For example, the theory of stationary motions of the gyroscope [1] and the more general theory of stationary solutions of dynamical systems on LIE groups suggests us to search the steady-deformation solutions of our problem:

$$D = \text{const}\,, \qquad X = \text{const.}\,, \qquad \Theta = \text{const}\,. \tag{3.16}$$

Substituting this into the equations of motion one finds that X, Θ are collinear ($X = \lambda\Theta$ for some real number λ) and each of them possesses only one non-trivial component, α, β, respectively. The equations of motion imply some algebraic conditions imposed on the constants $D_1, \dots, D_n, \alpha, \beta$. The full solution for the time-dependent matrix $\varphi(t)$ has then the following form:

$$\varphi(t) = e^{Xt}\, D\, e^{-\Theta t}\,. \tag{3.17}$$

In the special case $n = 3$, such solutions are related to the ellipsoidal figures of equilibrium studied in astrophysics [4].

4. Hamiltonian Description

We present here briefly the basic ideas of the canonical formalism for doubly-isotropic models.

Let us assume that the equations of motion are derivable from the Lagrangean $L = T - V$ with velocity-independent potential energy V.

We describe the LEGENDRE transformation directly in terms of non-holonomic objects. Varying the quantities X, Θ, D: $X \mapsto X + \delta X$, $\Theta \mapsto \Theta + \delta\Theta$, $D \mapsto D + \delta D$, we find the linear part of variation of the kinetic energy to be

$$\delta T = \sum_i p_i \delta \dot{D}_i + \tfrac{1}{2} \operatorname{Tr}(\pi_\Theta^T \delta\Theta) + \tfrac{1}{2} \operatorname{Tr}(\pi_X^T \delta X) \tag{4.1}$$

where

$$p_i = \frac{\partial T}{\partial \dot{D}_i} = \mu \frac{d}{dt} D_i$$

and the skew-symmetric matrices π_X, π_Θ are given by the following formulas:

$$\pi_X = \mu(D^2X + XD^2 - 2D\Theta D)\,, \qquad \pi_\Theta = \mu(D^2\Theta + \Theta D^2 - 2DXD)\,. \tag{4.2}$$

π_X, π_Θ are non-holonomic momenta conjugate to the angular velocities X, Θ, respectively.

The matrix elements of the matrices π_X, π_Θ are Hamiltonian generators of the following local LIE groups of transformations:

$$\pi_X\colon \varphi = ODR^T \mapsto OUDR^T\,; \qquad \pi_\Theta\colon \varphi = ODR^T \mapsto OD(RU)^T = ODU^TR^T \tag{4.3}$$

where $U \in SO(n, R)$.

Hence, the POISSON brackets of π_X have the structure of the n-dimensional orthogonal LIE algebra (algebra of skew-symmetric $n \times n$ matrices); the same concerns π_Θ. POISSON brackets of π_X with π_Θ vanish identically.

Transforming quantities π_X, π_Θ according to the adjoint rule: $\pi_X \mapsto O\pi_XO^T$, $\pi_\Theta \mapsto R\pi_\Theta R^T$, we obtain quantities obeying the same LIE algebra. Their matrix elements are Hamiltonian generators [16] of the left and right (space and material) rigid rotations:

$$\pi_X\colon \varphi = ODR^T \mapsto UODR^T = U\varphi\,, \qquad \pi_\Theta\colon \varphi = ODR^T \mapsto OD(UR)^T = ODR^TU^T = \varphi U^T \tag{4.4}$$

where U runs over the group $SO(n, R)$.

Therefore, when expressed by X, Θ, $\mathrm{d}D/\mathrm{d}t$, the quantity $O\pi_X O^T$ should equal the spin. The straightforward calculations show in fact that

$$S = O\pi_X O^T\,. \tag{4.5}$$

In Hamiltonian formalism all physical quantities, in particular the energy are expressed in terms of canonical momenta and generalized coordinates. Hence, when starting with Lagrangean description based on velocities, one has to invert the LEGENDRE transformation and express generalized velocities through canonical momenta (and coordinates). Unfortunately, for general n there is no compact matrix formula for inverting (4.2). However, one can easily describe the reciprocal LEGENDRE transformation for „physical" values of n, i.e. 3,2. Let us describe briefly the structure of the three-dimensional case, especially important in applications. Two-dimensional models will be treated separately in the next section.

We denote:

$$X = \begin{bmatrix} 0 & -X_3 & X_2 \\ X_3 & 0 & -X_1 \\ -X_2 & X_1 & 0 \end{bmatrix}, \qquad \Theta = \begin{bmatrix} 0 & -\Theta_3 & \Theta_2 \\ \Theta_3 & 0 & -\Theta_1 \\ -\Theta_2 & \Theta_1 & 0 \end{bmatrix}, \tag{4.6}$$

$$\pi_X = \begin{bmatrix} 0 & \pi_{X3} & -\pi_{X2} \\ -\pi_{X3} & 0 & \pi_{X1} \\ \pi_{X2} & -\pi_{X1} & 0 \end{bmatrix}, \qquad \pi_\Theta = \begin{bmatrix} 0 & \pi_{\Theta 3} & -\pi_{\Theta 2} \\ -\pi_{\Theta 3} & 0 & \pi_{\Theta 1} \\ \pi_{\Theta 2} & -\pi_{\Theta 1} & 0 \end{bmatrix} \tag{4.7}$$

and, obviously

$$D = \begin{bmatrix} D_1 & 0 & 0 \\ 0 & D_2 & 0 \\ 0 & 0 & D_3 \end{bmatrix}. \tag{4.8}$$

Then

$$T = \frac{\mu}{2}\sum_{i=1}^{3} A_i X_i^2 + \frac{\mu}{2}\sum_{i=1}^{3} A_i \Theta_i^2 - 2\mu \sum_{i=1}^{2} B_i X_i \Theta_i + \frac{\mu}{2}\sum_{i=1}^{3}\left(\frac{\mathrm{d}D_i}{\mathrm{d}t}\right)^2 \tag{4.9}$$

where

$$A_i = D_j^2 + D_k^2\,, \qquad i \neq j \neq k \neq i; \qquad B_i = D_j D_k\,, \quad i \neq j \neq k \neq i\,. \tag{4.10}$$

The LEGENDRE transformation takes on the following form:

$$p_i = \mu \frac{\mathrm{d}D_i}{\mathrm{d}t}, \qquad \pi_{Xi} = \mu A_i X_i - 2\mu B_i \Theta_i\,, \qquad \pi_{\Theta i} = -2\mu B_i X_i + \mu A_i \Theta_i\,. \tag{4.11}$$

It is easy to calculate the reciprocal mapping:

$$\frac{\mathrm{d}D_i}{\mathrm{d}t} = \frac{1}{\mu} p_i\,, \qquad X_i = \frac{1}{\mu} E_i \pi_{Xi} + \frac{2}{\mu} F_i \pi_{\Theta i}\,, \qquad \Theta_i = \frac{2}{\mu} F_i \pi_{Xi} + \frac{1}{\mu} E_i \pi_{\Theta i} \tag{4.12}$$

where

$$E_i = \frac{D_j^2 + D_k^2}{(D_j^2 - D_k^2)^2}\,, \qquad F_i = \frac{D_j D_k}{(D_j^2 - D_k^2)^2}\,, \qquad i \neq j \neq k \neq i\,. \tag{4.13}$$

When expressed in canonical terms, the kinetic energy becomes

$$T = \frac{1}{2\mu}\sum_{i=1}^{3} E_i \pi_{Xi}^2 + \frac{1}{2\mu}\sum_{i=1}^{3} E_i \pi_{\Theta i}^2 + \frac{2}{\mu}\sum_{i=1}^{3} F_i \pi_{Xi}\pi_{\Theta i} + \frac{1}{2\mu}\sum_{i=1}^{3} p_i^2\,. \tag{4.14}$$

The Hamiltonian function (total energy) of the doubly-isotropic potential model has the form:

$$H = T + V(D_1, D_2, D_3) \tag{4.15}$$

where V is invariant under permutations of its arguments. Similarly, as in Newtonian description the canonical equations of motion lead to the closed subsystem of equations imposed on parameters D_i, π_{Xi}, $\pi_{\Theta i}$:

$$\begin{gathered} \frac{\mathrm{d}}{\mathrm{d}t}\pi_{Xi} = \{\pi_{Xi}, H\}\,, \qquad \frac{\mathrm{d}}{\mathrm{d}t}\pi_{\Theta i} = \{\pi_{\Theta i}, H\}\,, \\ \frac{\mathrm{d}D_i}{\mathrm{d}t} = \{D_i, H\}\,, \qquad \frac{\mathrm{d}p}{\mathrm{d}t} = \{p_i, H\}\,, \quad \text{i.e.} \quad \mu\frac{\mathrm{d}^2 D_i}{\mathrm{d}t^2} = -\frac{\partial V}{\partial D_i}. \end{gathered} \tag{4.16}$$

POISSON brackets on the right hand sides can be easily calculated by help of the standard rules:

$$\{f, g\} = -\{g, f\}\,, \qquad \{\{f, g\}, h\} + \{\{g, h\}, f\} + \{\{h, f\}, g\} = 0\,, \qquad \{f, F(g)\} = \{f, g\}\frac{\mathrm{d}F}{\mathrm{d}g}. \tag{4.17}$$

One should substitute here the following basic formulas:

$$\begin{gathered} \{D_i, p_j\} = \delta_{ij}\,, \qquad \{\pi_{\Theta i}, \pi_{\Theta j}\} = \varepsilon_{ijk}\pi_{\Theta k}\,, \qquad \{\pi_{Xi}, \pi_{Xj}\} = \varepsilon_{ijk}\pi_{Xk}\,, \\ \{\pi_{Xi}, \pi_{\Theta j}\} = \{\pi_{Xi}, D_j\} = \{\pi_{Xi}, p_j\} = \{\pi_{\Theta i}, D_j\} = \{\pi_{\Theta i}, p_j\} = 0\,. \end{gathered} \tag{4.18}$$

The dynamical system (4.16) can be solved or at least studied qualitatively in the simplified phase space of variables $(D_i, P_j, \pi_{Xk}, \pi_{\Theta l})$. For example, similarly as in Newtonian theory, one can investigate special stationary solutions $(D_i, \pi_X, \pi_\Theta = \text{const.})$ and small vibrations about such solutions.

When some solution of (4.16) is known (rigorously, or in some approximation), then the time dependence of velocities can be algebraically calculated from (4.12). Then, one can calculate the time evolution of Green and Cauchy gyroscopes, $R(t)$, $O(t)$, by solving first-order differential equations (3.14). The final solution is $\varphi(t) = O(t)\, D(t)\, R^T(t)$.

Unfortunately, it turns out that in three-dimensional problems with reasonable, physically justified potentials it is impossible to solve the equations of motion by help of separation of variables techniques.

5. Integrable Two-Dimensional Problems

The mechanics of the two-dimensional (plane) affinely-rigid body is relatively simple when compared with the three-dimensional theory, both from the theoretical and computational point of view. The main reason is that the structural properties of orthogonal groups in two and three dimensions are quite different (even opposite): $SO(3, R)$ is a three-dimensional simple Lie group and $SO(2, R)$ is a one-dimensional and consequently commutative group of plane rotations. The simplicity of $SO(3, R)$ implies that in all formulas for important physical quantities (e.g. kinetic energy) and in equations of motion, the strain invariants D_1, D_2, D_3 are mixed in an essential, non-separable way. It is impossible to solve the classical Hamilton-Jacobi equation with a realistic, isotropic potential by means of the separation of variables method. Contrary, in the two-dimensional theory there exists a relatively wide class of isotropic potentials which admit rigorous, finite solutions. The corresponding Hamilton-Jacobi equation is separable and the formulas for the general shape of integrable potentials follow from the classical Stäckel theorem. Rigorous solutions for two-dimensional problems may be useful in the theory of molecular vibrations for plane molecules (like S_8, C_6H_6) and their aggregates (molecular crystals, microelastic continua). Perhaps, one could also use them in fluid dynamics [3]. Of course, the applications in macroscopic two-dimensional elastic problems are also possible (e.g. cylinders with homogeneously-deformable cross-sections).

Let us use the adapted description:

$$\varphi = ODR^T \tag{5.1}$$

where

$$O = \begin{bmatrix} \cos\vartheta & \sin\vartheta \\ -\sin\vartheta & \cos\vartheta \end{bmatrix}, \qquad D = \begin{bmatrix} D_1 & 0 \\ 0 & D_2 \end{bmatrix}, \qquad R = \begin{bmatrix} \cos\psi & \sin\psi \\ -\sin\psi & \cos\psi \end{bmatrix}. \tag{5.2}$$

The isotropic kinetic energy has the form:

$$T = \frac{\mu}{2}\left[(D_1^2 + D_2^2)\left(\frac{\mathrm{d}\vartheta}{\mathrm{d}t}\right)^2 + (D_1^2 + D_2^2)\left(\frac{\mathrm{d}\psi}{\mathrm{d}t}\right)^2 - 4D_1D_2\frac{\mathrm{d}\vartheta}{\mathrm{d}t}\frac{\mathrm{d}\psi}{\mathrm{d}t} + \left(\frac{\mathrm{d}D_1}{\mathrm{d}t}\right)^2 + \left(\frac{\mathrm{d}D_2}{\mathrm{d}t}\right)^2\right]. \tag{5.3}$$

The matrices O, R do not enter into this equation, hence the angles ϑ, ψ are cyclic variables. In coordinates $(\vartheta, \psi, D_1, D_2)$ the Hamilton-Jacobi equation is non-separable even in the interaction-free case. Let us introduce new coordinates η, γ, α, β given by the following formulas:

$$\alpha = \frac{D_1 + D_2}{\sqrt{2}}, \quad \beta = \frac{D_1 - D_2}{\sqrt{2}}, \qquad \eta = \vartheta - \psi, \qquad \gamma = \vartheta + \psi. \tag{5.4}$$

The kinetic energy becomes then:

$$T = \frac{\mu}{2}\left[\alpha^2\left(\frac{\mathrm{d}\eta}{\mathrm{d}t}\right)^2 + \beta^2\left(\frac{\mathrm{d}\gamma}{\mathrm{d}t}\right)^2 + \left(\frac{\mathrm{d}\alpha}{\mathrm{d}t}\right)^2 + \left(\frac{\mathrm{d}\beta}{\mathrm{d}t}\right)^2\right]. \tag{5.5}$$

This form is both diagonal and separable. Legendre transformation leads to the form:

$$T = \frac{1}{2\mu}\left[\frac{p_\eta^2}{\alpha^2} + \frac{p_\gamma^2}{\beta^2} + p_\alpha^2 + p_\beta^2\right] \tag{5.6}$$

where, obviously, p_η, p_γ, p_α, p_β are canonical momenta conjugate to η, γ, α, β, respectively. In isotropic dynamical problems, when

$$H = T + V(\alpha, \beta) \tag{5.7}$$

the angular canonical momenta p_η, p_γ are constants of motion (η, γ are cyclic variables). This enables us to reduce the number of effective degrees of freedom from four variables η, γ, α, β to the two ones α, β. For arbitrary but fixed constant values of the angular momenta, $p_\eta = A$, $p_\gamma = B$, the time-dependence of invariants α, β is governed by the following effective Hamiltonian H_{AB} in two degrees of freedom:

$$H_{A,B} = \frac{1}{2\mu}(p_\alpha^2 + p_\beta^2) + V(\alpha, \beta) + \frac{A^2}{2\mu}\frac{1}{\alpha^2} + \frac{B^2}{2\mu}\frac{1}{\beta^2} = \frac{1}{2\mu}(p_\alpha^2 + p_\beta^2) + V_{AB}(\alpha, \beta). \tag{5.8}$$

Formally, this Hamiltonian and the resulting equations of motion are isomorphic to those for the material point moving on a plane under the influence of the effective potential energy

$$V_{AB}(\alpha, \beta) = V(\alpha, \beta) + \frac{A^2}{2\mu}\frac{1}{\alpha^2} + \frac{B^2}{2\mu}\frac{1}{\beta^2}. \tag{5.9}$$

Hence, it is natural to expect that the well-known orthogonal curvilinear coordinates on the Euclidean (α, β)-plane will be especially useful when trying to solve the resulting Hamilton-Jacobi equation by means of the separation of variables. Therefore, we will try to use the following coordinates on the invariants-plane:

1. Cartesian coordinates α, β

2. Polar coordinates r, φ

defined in such a way that

$$\alpha = \sqrt{r} \cos \frac{\varphi}{2}, \qquad \beta = \sqrt{r} \sin \frac{\varphi}{2}. \tag{5.10}$$

Hence, strictly speaking, polar coordinates on the (α, β)-plane are given by $\sqrt{r}$ and $\varphi/2$. However, the quantities r, φ are more convenient in calculations.

3. Elliptic coordinates ϱ, λ:

$$\alpha = \sqrt{2} \operatorname{ch} \varrho \cos \lambda, \qquad \beta = \sqrt{2} \operatorname{sh} \varrho \sin \lambda, \tag{5.11}$$

the factor $\sqrt{2}$ being chosen for the sake of convenience (the natural configuration $\varphi = I$ corresponds then to the nice values of parameters: $\varrho = 0, \lambda = 0$).

4. Parabolic coordinates ξ, η:

$$\alpha = \tfrac{1}{2}(\eta^2 - \xi^2), \qquad \beta = \xi\eta. \tag{5.12}$$

5. Bi-polar coordinates x, y:

$$\alpha = \frac{a \operatorname{sh} x}{\operatorname{ch} x - \cos y}, \qquad \beta = \frac{a \sin y}{\operatorname{ch} x - \cos y}. \tag{5.13}$$

It turns out that parabolic and bi-polar coordinates have to be rejected, because the resulting Hamilton-Jacobi equations are non-separable even in the interaction-free case $V = 0$. This is because of the terms $\frac{1}{2\mu} \frac{p_\eta^2}{\alpha^2}, \frac{p_\gamma^2}{\beta^2} \frac{1}{2\mu}$ in H $\left(\text{or equivalently-the ,,centrifugal" terms in the reduced potential } V_{AB} \left/ \frac{A^2}{2\mu} \frac{1}{\alpha^2}, \frac{B^2}{2\mu} \frac{1}{\beta^2} \right)\right.$. On the contrary, the cartesian, polar and elliptic coordinates enable us to employ the separation of variables technique in the interaction-free case. The kinetic energy takes the following forms in polar and elliptic coordinates:

$$T = \frac{\mu}{2} \left[r \cos^2 \frac{\varphi}{2} \left(\frac{d\eta}{dt}\right)^2 + r \sin^2 \frac{\varphi}{2} \left(\frac{d\gamma}{dt}\right)^2 + \frac{1}{4r} \left(\frac{dr}{dt}\right)^2 + \frac{r}{4} \left(\frac{d\varphi}{dt}\right)^2 \right], \tag{5.14}$$

$$T = \mu \left[\operatorname{ch}^2 \varrho \cos^2 \lambda \left(\frac{d\eta}{dt}\right)^2 + \operatorname{sh}^2 \varrho \sin^2 \lambda \left(\frac{d\gamma}{dt}\right)^2 + (\operatorname{ch}^2 \varrho - \cos^2 \lambda) \left(\frac{d\varrho}{dt}\right)^2 + (\operatorname{ch}^2 \varrho - \cos^2 \lambda) \left(\frac{d\lambda}{dt}\right)^2 \right]. \tag{5.15}$$

The classical Stäckel theorem leads to the following general classes of integrable (separable) potentials corresponding to the cartesian, polar and elliptic variables:

$$V(\vartheta, \psi, \alpha, \beta) = \frac{V_\eta(\vartheta - \psi)}{\alpha^2} + \frac{V_\gamma(\vartheta + \psi)}{\beta^2} + V_\alpha(\alpha) + V_\beta(\beta), \tag{5.16}$$

$$V(\vartheta, \psi, r, \varphi) = \frac{V_\eta(\vartheta - \psi)}{r \cos^2 \frac{\varphi}{2}} + \frac{V_\gamma(\vartheta + \psi)}{r \sin^2 \frac{\varphi}{2}} + V_r(r) + \frac{V_\varphi(\varphi)}{r}, \tag{5.17}$$

$$V(\vartheta, \psi, \varrho, \lambda) = \frac{V_\eta(\vartheta - \psi)}{2 \operatorname{ch}^2 \varrho \cos^2 \lambda} + \frac{V_\gamma(\vartheta + \psi)}{2 \operatorname{sh}^2 \varrho \sin^2 \lambda} + \frac{V_\varrho(\varrho)}{2(\operatorname{ch}^2 \varrho - \cos^2 \lambda)} + \frac{V_\lambda(\lambda)}{2(\operatorname{ch}^2 \varrho - \cos^2 \lambda)}. \tag{5.18}$$

In these formulas, V_η, V_γ, V_α, V_β, V_r, V_φ, V_ϱ, V_λ are arbitrary (but regular enough) functions of a single variable (indicated as an argument).

We are especially interested in doubly-isotropic models, when the potential energy does not depend on the angular variables ϑ, ψ (equivalently — η, γ). The most general isotropic forms of the "cartesian" and "polar" potentials result when one puts $V_\eta = O$, $V_\gamma = 0$ in formulas (5.16) (5.17). Such a restriction would be too strong in the "elliptic" case — the most general isotropic form of (5.18) correspondes to $V_\eta = \text{const}$, $V_\gamma = \text{const}$. There is no reason to admit non-zero constant values of V_η, V_γ in (5.16), (5.17), because this would be exactly equivalent to modifying arbitrary functions V_α, V_β, V_φ.

Finally, the most general, isotropic, separable potentials corresponding to our choice of coordinate systems, have the following shape:

$$V(\alpha, \beta) = V_\alpha(\alpha) + V_\beta(\beta), \qquad V(r, \varphi) = V_r(r) + \frac{V_\varphi(\varphi)}{r}, \tag{5.19}$$

$$V(\varrho, \lambda) = \frac{A}{2 \operatorname{ch}^2 \varrho \cos^2 \lambda} + \frac{B}{2 \operatorname{sh}^2 \varrho \sin^2 \lambda} + \frac{V_\varrho(\varrho)}{2(\operatorname{ch}^2 \varrho - \cos^2 \lambda)} + \frac{V_\lambda(\lambda)}{2(\operatorname{ch}^2 \varrho - \cos^2 \lambda)}$$

(A, B — constants).

The question arises for the existence of potentials which could lead to Hamilton-Jacobi equations separable simultaneously in all three aforementioned coordinate systems. It is well known from analytical mechanics that such a simultaneous separability leads to some extra degeneracy of the problem (additional constants of motion). One can easily show that the only potentials with the above property are those given by the formula

$$V = \frac{A}{\alpha^2} + \frac{B}{\beta^2} + C(\alpha^2 + \beta^2) = \frac{2A}{(D_1 + D_2)^2} + \frac{2B}{(D_1 - D_2)^2} + C(D_1^2 + D_2^2) \tag{5.20}$$

where A, B, C are constants. Of course, we mean here only doubly-isotropic potentials.

Among the aforementioned three classes of isotropic separable potentials it is only that of polar potentials $V = V_r(r) + \frac{V_\varphi(\varphi)}{r}$ what has a chance to be entirely compatible with standard requirements of macroscopic, continuous elasticity. In fact, by the appropriate choice of arbitrary functions V_r, V_φ one can achieve both the stable equilibrium (minimum of V) in the natural state (when $D_1 = D_2 = 1$, i.e. $r = 2$, $\varphi = 0$) and the invariance of the physical region of parameters ($D_1 > 0$, $D_2 > 0$) under the one-parameter dynamical group generated by the Hamiltonian $H = T + V$ (i.e. compatibility of inequalities $D_1 > 0$, $D_2 > 0$ with equations of motion). In particular, this prevents the body from the "collapsed" configurations in which $\det \varphi = D_1 D_2 = 0$. The algebraically simplest model of such a potential has the form:

$$V = \frac{2\varkappa}{r \cos \varphi} + \frac{\varkappa}{2} r = \varkappa \left(\frac{1}{D_1 D_2} + \frac{D_1^2 + D_2^2}{2} \right). \tag{5.21}$$

There exists an infinity of polar potentials with the aforementioned properties.

On the contrary, all "cartesian" and "elliptic" separable potentials are incompatible with macroscopic physical restrictions $D_1 > 0$, $D_2 > 0$. With such potentials it is always possible to find solutions (trajectories) changing the sign of D_i. After some time such trajectories leave the physical region $D_1 > 0$, $D_2 > 0$. Of course, this is not a serious problem when investigating deformations of discrete plane systems (e.g. a finite system of material points on a plane). However, in macroscopic continuum theory the "collapsing" trajectories crossing the singular surface $\det \varphi = D_1 D_2 = 0$ are unacceptable.

As we know from analytical mechanics, the separability of the Hamilton-Jacobi equation in some coordinates $(q^1, \ldots, q^n)$ (n is the number of degrees of freedom) implies the existence of n constants of motion $F_1, \ldots, F_n$ with pairwise vanishing Poisson brackets: $\{F_i, F_j\} = 0$. Now, we present the general form of those constants of motion and the corresponding complete integrals of the Hamilton-Jacobi equation for our doubly-isotropic models:

I. Cartesian coordinates α, β:

$$V(\alpha, \beta) = V_\alpha(\alpha) + V_\beta(\beta)\,. \tag{5.22}$$

Canonical momenta conjugate to α, β, ϑ, ψ are denoted as p_α, p_β, p_ϑ, p_ψ.
Constants of motion:

$$1.\ p_\vartheta\,, \quad 2.\ p_\psi\,, \quad 3.\ H_\alpha = \frac{1}{2\mu}\left(\frac{(p_\vartheta - p_\psi)^2}{4\alpha^2} + p_\alpha^2\right) + V_\alpha\,, \quad 4.\ H_\beta = \frac{1}{2\mu}\left(\frac{(p_\vartheta + p_\psi)^2}{4\beta^2} + p_\beta^2\right) + V_\beta\,. \tag{5.23}$$

The total energy $H = H_\alpha + H_\beta$ is a constant of motion functionally depending on the above ones.
Complete integral:

$$S(\alpha, \beta, \vartheta, \psi, t; E_\alpha, E_\beta, a, b) = (E_\alpha + E_\beta)\, t + a\vartheta + b\psi +$$
$$+ \int \sqrt{2\mu(E_\alpha - V_\alpha) - \frac{(a-b)^2}{4\alpha^2}}\, d\alpha + \int \sqrt{2\mu(E_\beta - V_\beta) - \frac{(a+b)^2}{4\beta^2}}\, d\beta\,, \tag{5.24}$$

E_α, E_β, a, b being arbitrary constants.

II. Polar coordinates r, φ:

$$V(r, \varphi) = V_r(r) + V_\varphi(\varphi)/r\,. \tag{5.25}$$

When expressed in terms of canonical momenta p_r, p_φ, p_ϑ, p_ψ, our constants of motion take the form:

$$1.\ p_\vartheta\,, \quad 2.\ p_\psi\,, \quad 3.\ h_\varphi = \frac{1}{2\mu}\left(\frac{p_\vartheta^2 + p_\psi^2 + 2p_\vartheta p_\psi \cos\varphi}{\sin^2\varphi} + 4p_\varphi^2\right) + V_\varphi\,,$$

4. The total energy $H = T + V = H_r + h_\varphi/r$ where $H_r = (2r/\mu)\, p_r^2 + V_r\,.$ (5.26)

Complete integral:

$$S(r, \varphi, \vartheta, \psi, t; E, e_\varphi, a, b) = Et + a\vartheta + b\psi +$$
$$+ \int \sqrt{\frac{\mu}{2r}(E - V_r) - \frac{\mu e_\varphi}{2r^2}}\, dr + \int \sqrt{\frac{\mu}{2}(e_\varphi - V_\varphi) - \frac{a^2 + b^2 + 2ab\cos\varphi}{4\sin^2\varphi}}\, d\varphi\,, \tag{5.27}$$

E, e_φ, a, b — constants.

III. Elliptic coordinates ϱ, λ

$$V = \frac{A}{2\,\mathrm{ch}^2\varrho\cos^2\lambda} + \frac{B}{2\,\mathrm{sh}^2\varrho\sin^2\lambda} + \frac{V_\varrho(\varrho)}{2(\mathrm{ch}^2\varrho - \cos^2\lambda)} + \frac{V_\lambda(\lambda)}{2(\mathrm{ch}^2\varrho - \cos^2\lambda)}. \tag{5.28}$$

We use canonical momenta $p_\varrho, p_\lambda, p_\vartheta, p_\psi$. Before writing explicitly our constants of motion we introduce some auxiliary quantities:

$$h_\varrho = \frac{1}{2\mu}\left[p_\varrho^2 + 2\mu V_\varrho - \frac{\frac{1}{4}(p_\vartheta - p_\psi)^2 + 2A\mu}{\mathrm{ch}^2\varrho} + \frac{\frac{1}{4}(p_\vartheta + p_\psi)^2 + 2B\mu}{\mathrm{sh}^2\varrho}\right], \tag{5.29}$$

$$h_\lambda = \frac{1}{2\mu}\left[p_\lambda^2 + 2\mu V_\lambda + \frac{\frac{1}{4}(p_\vartheta - p_\psi)^2 + 2A\mu}{\cos^2\lambda} + \frac{\frac{1}{4}(p_\vartheta + p_\psi)^2 + 2B\mu}{\sin^2\lambda}\right]. \tag{5.30}$$

Constants of motion:

$$1.\ p_\vartheta, \qquad 2.\ p_\psi, \qquad 3.\ \mathcal{K} = \frac{h_\varrho\cos^2\lambda + h_\lambda\,\mathrm{ch}^2\varrho}{2(\mathrm{ch}^2\varrho - \cos^2\lambda)}, \qquad 4.\ \mathcal{L} = \frac{h_\varrho\sin^2\lambda - h_\lambda\,\mathrm{sh}^2\varrho}{2(\mathrm{sh}^2\varrho + \sin^2\lambda)}. \tag{5.31}$$

The total energy $H = T + V$ is a constant of motion depending on the above ones:

$$H = \mathcal{K} + \mathcal{L}. \tag{5.32}$$

Complete integral:

$$\begin{aligned} S(\varrho, \lambda, \vartheta, \psi, t; E, \varkappa, a, b) &= Et + a\vartheta + b\psi + \\ &+ \int\sqrt{4\mu(E\,\mathrm{ch}^2\varrho - \varkappa) - 2\mu V_\varrho - \frac{(a+b)^2 + 8\mu B}{4\,\mathrm{sh}^2\varrho} + \frac{(a-b)^2 + 8\mu A}{4\,\mathrm{ch}^2\varrho}}\,\mathrm{d}\varrho + \\ &+ \int\sqrt{4\mu(\varkappa - E\cos^2\lambda) - 2\mu V_\lambda - \frac{(a+b)^2 + 8\mu B}{4\sin^2\lambda} - \frac{(a-b)^2 + 8\mu A}{4\cos^2\lambda}}\,\mathrm{d}\lambda. \end{aligned} \tag{5.33}$$

According to the general rules of analytical mechanics the complete integral is equivalent to the general solution of equations of motion. Namely, the complete integral $S(q^i, t; a^i)$ leads to the following finite equations for trajectories $\partial S/\partial a^i = b^i$, $i = 1, \ldots, n$, where $(a^1, \ldots, a^n; b^1, \ldots, b^n)$ are constants parametrizing the general solution.

Hence, for our class of dynamical models of a plane affinely-rigid body, the deformative behaviour of a system can be found by means of calculating one-dimensional integrals.

References

1 Arnold, V. I., Matematicheskie metody klassicheskoi mekhaniki, (in Russian), Nauka, Moscow 1974.
2 Bogoyavlenskij, O. I., Kolebatelnyj regim rasširenia gazowogo oblaka w wakuum, Piśma w astronomičeskij jurnal, **1**, nr. 9, 1975 (in Russian).
3 Bogoyavlenskij, O. I., Dinamika gravitirujuščego gazowogo elipsoida, Prikladnaja matematika i mekhanika **40**, nr. 2, 1976.
4 Chandrasekhar, S., Ellipsoidal Figures of Equilibrium, New Haven and London, Yale University Press 1969.
5 Dyson, F. J., Dynamics of a Spinning Gas Cloud, Journal of Mathematics and Mechanics, **18**, nr. 1, 1968.
6 Eringen, C. E., Mechanics of Micromorphic Continua, Proc. of the IUTAM-Symposium, Freudenstadt and Stuttgart 1967, 18, Springer, Berlin-Heidelberg-New York 1968, Ed. E. Kröner.
7 Hehl, F. W., On Hypermomentum in General Relativity, I. The Notion of Hypermomentum, Zeitschrift für Naturforschung **31a**, 1976.
8 Hehl, F. W.; Lord, E. A.; Ne'eman, Y., Hadron Dilation, Shear and Spin as Components of the Intrinsic Hypermomentum Current and Metric-Affine Theory of Gravitation, Physics Letters **71B**, nr. 2, 1977.
9 Landau, L. D.; Lifšic, E. M., Mekhanika Splošnych sred, Moskwa 1954.
10 Römer, H.; Westpfahl, K., Rotator-Spinteilchen in schwachen Gravitationsfeldern, Annalen der Physik, **22**, nr. 5/6, 1969.
11 Sławianowski, J. J., Analytical Mechanics of Finite Homogeneous Strains, Archives of Mechanics, **26**, nr. 4, 1974.
12 Sławianowski, J. J., Newtonian Dynamics of Homogeneous Strains, Archives of Mechanics, **27**, nr. 1, 1975.
13 Sławianowski, J. J., The Mechanics of an Affinely-Rigid Body, International Journal of Theoretical Physics, **12**, nr. 4, 1975.
14 Sławianowski, J. J., Newtonian Dynamics of Polynomial Deformations, Bull. de l'Academie Polonaise des Sciences (Serie des sciences techniques, XXIII, nr. 11, 1975).
15 Sławianowski, J. J., Deformable Gyroscope in a Non-Euclidean Space, Classical, Non-Relativistic Theory, Reports on Mathematical Physics, **10**, nr. 2, 1976.
16 Sternberg, S., Lectures on Differential Geometry, Prentice Hall, Inc. Englewood Cliffs, 1964.

Eingereicht am 4. 7. 1980

Anschrift: Doz. Dr. Jan J. Sławianowski, Polish Academy of Sciences, Institute of Fundamental Technological Research, Świętokrzyska 21, 00—049 Warsaw, Poland

ZAMM 62, 241 – 247 (1982)

H. BALKE

Zum Spannungsverzerrungsgesetz bei hyper- und hypoelastischen Verzweigungsproblemen

Isotrope hyper- und hypoelastische Elastizitätsgesetze werden in ihrer Beziehung zum klassischen Verzweigungsproblem diskutiert. Die in der Verzweigungstheorie übliche Vernachlässigung der Grunddeformation bedingt bei Kompressibilität des Materials die Gleichwertigkeit verschiedener hyper- und hypoelastischer Ansätze. Im inkompressiblen Fall und bei Anwesenheit hoher hydrostatischer Drücke zeigen die Deformationszuwüchse hyper- und hypoelastischer Körper unterschiedliches Verzweigungsverhalten. Dies wird am Beispiel einer Scherbandbildung gezeigt.

Isotropic hyperelastic and hypoelastic constitutive equations are investigated in relation to the classical bifurcation problem. The conventional assumption is made that the ground deformation is dropped in the incremental constitutive equations deduced from the finite hyperelastic relations. These equations are equivalent to some hypoelastic equations given a priori. If the materials are incompressible and undergo high hydrostatic pressure, then the velocity fields of the incremental hyperelastic and hypoelastic equations differ under the conditions of bifurcation. This is demonstrated for a simple example of a shear band.

Исследуются изотропные гипер-и гипоупругие определяющие соотношения в связи с классической бифуркационной проблемой. Обычно пренебрегают основную деформацию в бифуркационной теории. То гипер- и гипоупругие определяющие соотношения для сжимаемых материалов эквивалентны. Но в отличии от этого поведение несжимаемых гипер- и гипоупругих материалов при высоком гидростатическом давлении различно, как показано на локализации деформации в форме ленты.

0. Problemstellung

In der klassischen elastischen Verzweigungstheorie wird bekanntlich die Eindeutigkeit der zu einer finiten Grunddeformation differentiell benachbarten Deformation geprüft. Damit die manchmal schwierige Bestimmung der finiten Grundlösung entfällt und die Lösung des Verzweigungsproblems sich vereinfacht, ist es wünschenswert, die Grunddeformation im Elastizitätsgesetz zu vernachlässigen. Die damit berechneten kritischen Spannungen können diese Voraussetzung verletzen, wenn sie von der Größenordnung der elastischen Moduli sind. Bei Benutzung eines von vornherein differentiellen (hypoelastischen) Elastizitätsgesetzes erscheint dieser Sachverhalt nicht deutlich, da die finite Deformation nicht explizit in den verwendeten Gleichungen auftritt, und vielleicht deshalb werden in der Literatur Verzweigungslösungen angegeben bzw. Lösungsregime diskutiert, für die die kritische Spannung von der Ordnung des Schubmoduls ist [1, 2], oder es wird die Anwendbarkeit der Theorie statt durch ihre Voraussetzungen mittels der Eigenschaft der Konstruktionsmaterialien eingegrenzt, daß deren ertragbare Spannungen klein gegenüber den elastischen Moduli sind (z. B. [3, 4]). Andererseits ist ein hypoelastischer Ansatz bei Verletzung o. g. Voraussetzung nicht willkürfrei in dem Sinne, daß die gegenüber den Termen von der Größenordnung elastischer Modul mal Deformationsgeschwindigkeit sonst vernachlässigbaren, nicht eindeutigen Terme der Größenordnung Spannung mal Deformationsgeschwindigkeit wesentlich werden [5]. Bei inkompressiblem Material entsteht deshalb die Frage, ob der isotrope Spannungsanteil zur Erfüllung der o. g. Voraussetzungen klein sein muß, da er die Grunddeformation nur bedingt beeinflußt. Und schließlich ist es von Interesse zu wissen, inwieweit im Rahmen der angesprochenen Materialmodelle bei Annahme kleiner Verzerrungen bzw. im Vergleich zum Schubmodul kleiner Deviatorspannungen die Verzweigungslösungen noch konstitutiv beeinflußbar sind.

In der vorliegenden Arbeit wird unter der Annahme isotropen Materialverhaltens ein finites, hyperelastisches Gesetz vorausgesetzt, um das benötigte differentielle Elastizitätsgesetz herzuleiten. Nach Vernachlässigung der Grunddeformation erfolgt der Vergleich mit speziellen hypoelastischen Ansätzen aus der Literatur unter Beachtung der Kompressibilität bzw. Inkompressibilität des Materials und zwar direkt und mit Bezug auf die klassischen Verzweigungsbedingungen.

Besondere Unterschiede der beiden Materialmodelle werden an einem einfachen Beispiel elastisch-inkompressibler Scherbandbildung diskutiert.

1. Kinematische und dynamische Grundlagen

Zunächst seien die benötigten Variablen [6, 7] in der Matrizennotation zusammengestellt. Halbfette, große lateinische Buchstaben bezeichnen Tensoren zweiter Stufe, halbfette, kleine lateinische Buchstaben Vektoren. Außerdem wird die kartesische Koordinatenschreibweise benutzt, wenn es zur Formulierung spezieller Gleichungen oder für Erläuterungen erforderlich ist.

Die Bewegung des betrachteten Körpers werde beschrieben durch die räumliche Lage $\boldsymbol{x}$ des typischen Teilchens in der Momentankonfiguration φ zur Zeit t, das in der Ausgangskonfiguration Φ die Position $\boldsymbol{X}$ besaß.

$$\boldsymbol{x} = \boldsymbol{x}(\boldsymbol{X}, t) , \qquad x_k = x_k(X_K, t) . \tag{1.1}$$

Der Deformationsgradient $\boldsymbol{F}$

$$\boldsymbol{F} = \frac{\partial \boldsymbol{x}}{\partial \boldsymbol{X}} , \qquad F_{kK} = \frac{\partial x_k}{\partial X_K} \tag{1.2}$$

läßt sich eindeutig zerlegen in den symmetrischen linken Strecktensor $\boldsymbol{V}$ und den Drehungstensor $\boldsymbol{R}$

$$\boldsymbol{F} = \boldsymbol{VR}\,, \qquad \boldsymbol{RR}^T = \boldsymbol{I}\,, \qquad \det \boldsymbol{R} = 1\,, \tag{1.3}$$

$\boldsymbol{R}^T = \boldsymbol{R}^{-1}$ — zu $\boldsymbol{R}$ transponierter Tensor, $\boldsymbol{I}$ — Einheitstensor. Die Hauptwerte von $\boldsymbol{V}$ sind

$$V_\varepsilon = \left(\frac{l}{L}\right)_\varepsilon, \qquad \varepsilon = 1, 2, 3\,, \tag{1.4}$$

wo l die momentane und L die Ausgangslänge der Schenkel eines jederzeit orthogonalen Dreibeins bezeichnen.

Der linke CAUCHY-GREEN-Tensor ist

$$\boldsymbol{B} = \boldsymbol{V}^2\,, \tag{1.5}$$

mit den durch seine Hauptwerte B_ε ausgedrückten Invarianten

$$\mathrm{I}_B = B_1 + B_2 + B_3\,, \qquad \mathrm{II}_B = B_1B_2 + B_1B_3 + B_2B_3\,, \qquad \mathrm{III}_B = B_1B_2B_3\,. \tag{1.6}$$

Aus der Geschwindigkeit

$$\boldsymbol{u} = \dot{\boldsymbol{x}} = \frac{\partial \boldsymbol{x}(\boldsymbol{X}, t)}{\partial t}, \tag{1.7}$$

folgen der Geschwindigkeitsgradient

$$\boldsymbol{L} = \frac{\partial \boldsymbol{u}}{\partial \boldsymbol{x}} = \frac{\partial \boldsymbol{u}}{\partial \boldsymbol{X}}\frac{\partial \boldsymbol{X}}{\partial \boldsymbol{x}} = \dot{\boldsymbol{F}}\boldsymbol{F}^{-1} \tag{1.8}$$

bzw. mit (1.3)

$$\boldsymbol{L} = \dot{\boldsymbol{V}}\boldsymbol{V}^{-1} + \boldsymbol{V}\dot{\boldsymbol{R}}\boldsymbol{R}^{-1}\boldsymbol{V}^{-1} \tag{1.9}$$

und seine Transponierte wegen

$$(\boldsymbol{R}^{-1})^{\cdot} = -\boldsymbol{R}^{-1}\dot{\boldsymbol{R}}\boldsymbol{R}^{-1}\,, \qquad \boldsymbol{L}^T = \boldsymbol{V}^{-1}\boldsymbol{V}^{\cdot} - \boldsymbol{V}^{-1}\dot{\boldsymbol{R}}\boldsymbol{R}^{-1}\boldsymbol{V}\,. \tag{1.10), (1.11}$$

Durch rechtsseitige Multiplikation von (1.9) mit $\boldsymbol{V}^2$, linksseitige Multiplikation von (1.11) mit $\boldsymbol{V}^2$ und anschließende Addition der beiden Gleichungen entsteht

$$\boldsymbol{LV}^2 + \boldsymbol{V}^2\boldsymbol{L}^T = \dot{\boldsymbol{V}}\boldsymbol{V} + \boldsymbol{V}\dot{\boldsymbol{V}} = \dot{\boldsymbol{B}}\,. \tag{1.12}$$

Der Geschwindigkeitsgradient läßt sich zerlegen in die symmetrische Deformationsgeschwindigkeit

$$\boldsymbol{D} = \tfrac{1}{2}\,(\boldsymbol{L} + \boldsymbol{L}^T) \tag{1.13}$$

und die antimetrische Drehgeschwindigkeit

$$\boldsymbol{W} = \tfrac{1}{2}\,(\boldsymbol{L} - \boldsymbol{L}^T)\,. \tag{1.14}$$

Der wahre Spannungstensor $\boldsymbol{T}$ ist als Summe seines isotropen Anteils und seines Deviators $\boldsymbol{S}$ darstellbar:

$$\boldsymbol{T} = \boldsymbol{S} - q\boldsymbol{I}\,, \qquad q = -\tfrac{1}{3}\,\mathrm{tr}\,\boldsymbol{T}\,, \tag{1.15}$$

$\mathrm{tr}\,\boldsymbol{T}$ — Spur von $\boldsymbol{T}$. Die Nennspannung $\boldsymbol{T}_n$ ist definiert durch

$$\boldsymbol{T}_n = \frac{\varrho_0}{\varrho}\,\boldsymbol{F}^{-1}\boldsymbol{T}\,, \tag{1.16}$$

ϱ, ϱ_0 — momentane bzw. Ausgangsmassendichte.

Die materielle Zeitableitung von (1.16) liefert mit [6]

$$\left(\frac{\varrho_0}{\varrho}\right)^{\cdot} = \frac{\varrho_0}{\varrho}\,\mathrm{tr}\,\boldsymbol{L}\,, \qquad \dot{\boldsymbol{T}}_n = \frac{\varrho_0}{\varrho}\,\boldsymbol{F}^{-1}(\boldsymbol{T}\,\mathrm{tr}\,\boldsymbol{L} - \boldsymbol{LT} + \dot{\boldsymbol{T}})\,. \tag{1.17), (1.18}$$

Wird die Momentankonfiguration φ gleich der Ausgangskonfiguration Φ gesetzt, so folgt aus (1.18) die Nennspannungsgeschwindigkeit im Grundzustand $\dot{\boldsymbol{T}}_G$:

$$(\dot{\boldsymbol{T}}_n)_{\varphi=\Phi} = \boldsymbol{T}\,\mathrm{tr}\,\boldsymbol{L} - \boldsymbol{LT} + \dot{\boldsymbol{T}} = \dot{\boldsymbol{T}}_G\,. \tag{1.19}$$

Bei Inkompressibilität gilt wegen (1.17)

$$\mathrm{tr}\,\boldsymbol{L} = 0\,. \tag{1.20}$$

2. Verzweigungstheorie

Zur Prüfung der Eindeutigkeit der Geschwindigkeitsfelder, die hier stellvertretend für Verschiebungszuwachsfelder stehen, dienen Kriterien, die von der Nennspannungsgeschwindigkeit im Grundzustand und der Geschwindigkeit abhängen.

Örtlich ausgedehnte Verzweigungslösungen im Grundzustand werden beschrieben [5, 7] durch die Gleichung des fortgesetzten Gleichgewichtes

$$\frac{\partial \dot{T}_{Gkl}}{\partial x_k} = 0\,, \tag{2.1}$$

im Inneren v des Körpers und die homogenen Randbedingungen auf der Oberfläche s

$$\dot{T}_{Gkl} n_k = 0 \,, \tag{2.2}$$

$$u_l = 0 \,, \tag{2.3}$$

($\boldsymbol{n}$ — äußere Flächeneinheitsnormale im Grundzustand, zeitlich konstant). Die Anzahl der Gleichungen (2.2), (2.3) ergänzt sich auf jedem Teil von s zu drei. In (2.2) bleiben die vorgegebenen Oberflächenkräfte während der inkrementellen Zusatzverformung richtungstreu.

Eine andere für den Eindeutigkeitsverlust der Geschwindigkeitsfelder notwendige Bedingung folgt aus (2.1), ..., (2.3) nach Kirchhoff [7]:

$$\int\limits_v \operatorname{tr} (\dot{\boldsymbol{T}}_G \boldsymbol{L}) \, dv = 0 \,. \tag{2.4}$$

(Das Ungleichheitszeichen anstelle des Gleichheitszeichens ist hinreichend für Eindeutigkeit.)

Verzweigungslösungen in bandförmigen Bereichen und das Kriterium für die Lösungsregime von (2.1) gehorchen der Gleichung [1, 2, 4]

$$\operatorname{tr} (\dot{\boldsymbol{T}}_G \boldsymbol{L}) = 0 \,, \tag{2.5}$$

wo der Geschwindigkeitsgradient sich darstellen läßt durch

$$L_{kl} = g_k e_l \,, \tag{2.6}$$

$\boldsymbol{g}$ — Normalenableitung der Geschwindigkeit, $\boldsymbol{e}$ — Einheitsvektor normal zum Band.

Für die Bildung der Nennspannungsgeschwindigkeit im Grundzustand wird nun ein differentielles Elastizitätsgesetz benötigt.

3. Elastisches Spannungsverzerrungsgesetz

In der Literatur zu Verzweigungsproblemen werden verschiedene hypoelastische Spannungsverzerrungsgesetze hauptsächlich zur Formulierung elastisch-plastischer Gleichungen angegeben, die sich in der Definition der Spannungsgeschwindigkeit unterscheiden. In [1, 3] wird die Jaumann-Ableitung der wahren Spannung benutzt, in [2, 8] die Jaumann-Ableitung der mitgeführten Kirchhoffschen Pseudospannung und in [8] außerdem die kovariante konvektive Ableitung der mitgeführten Kirchhoffschen Pseudospannung. In [4] erfolgt die Bezugnahme auf eine allgemeine Klasse von Spannungsgeschwindigkeiten gemäß [10]. Die allgemeinste in der Deformationsgeschwindigkeit lineare Definition dient in [9] zur Untersuchung spezieller Spannungszustände. Alle Definitionen unterscheiden sich, wie einleitend schon bemerkt, um Terme der Größenordnung Spannung mal Deformationsgeschwindigkeit.

Im folgenden wird die Spannungsgeschwindigkeit durch materielle Zeitableitung eines finiten isotropen hyperelastischen Ansatzes gebildet. Die Bewegung soll dabei in den Bestandteilen Deformationsgeschwindigkeit, Drehgeschwindigkeit und der anschließend zu vernachlässigenden Verzerrung dargestellt werden.

Das anzunehmende Verzerrungsenergiepotential Σ besitzt im allgemeinsten Fall die Form [7]

$$\Sigma = \Sigma(\mathrm{I}_B, \mathrm{II}_B, \mathrm{III}_B) \,. \tag{3.1}$$

Die Spannungsverzerrungsbeziehungen lauten damit [7]

$$\boldsymbol{T} = 2 \frac{\varrho}{\varrho_0} \left[\left(\mathrm{II}_B \frac{\partial \Sigma}{\partial \mathrm{II}_B} + \mathrm{III}_B \frac{\partial \Sigma}{\partial \mathrm{III}_B} \right) \boldsymbol{I} + \frac{\partial \Sigma}{\partial \mathrm{I}_B} \boldsymbol{B} - \mathrm{III}_B \frac{\partial \Sigma}{\partial \mathrm{II}_B} \boldsymbol{B}^{-1} \right] \tag{3.2}$$

bzw. für inkompressibles Material

$$\mathrm{III}_B = B_1 B_2 B_3 = 1, \qquad \boldsymbol{T} = \boldsymbol{S} - q\boldsymbol{I} = -p\boldsymbol{I} + 2 \frac{\partial \tilde{\Sigma}(\mathrm{I}_B, \mathrm{II}_B)}{\partial \mathrm{I}_B} \boldsymbol{B} - 2 \frac{\partial \tilde{\Sigma}(\mathrm{I}_B, \mathrm{II}_B)}{\partial \mathrm{II}_B} \boldsymbol{B}^{-1} \,, \qquad (3.3), (3.4)$$

p, $(p - q)$-skalare Größen, die sich aus dem jeweiligen Spannungszustand unter Berücksichtigung von (3.3) bestimmen.

Wegen der später zu vernachlässigenden kleinen Verzerrungen reichen anstelle von (3.2.), (3.4) z. B. folgende spezielle Approximationen aus:

$$\Sigma = 2G \left[\frac{1}{2} (\mathrm{I}_B - 3) + \frac{1}{8} \left(1 + \frac{\nu}{1 - 2\nu} \right) (\mathrm{I}_B - 3)^2 - \frac{1}{4} (\mathrm{II}_B - 3) \right], \qquad \tilde{\Sigma} = \frac{1}{4} G [\mathrm{I}_B - 3 + \mathrm{II}_B - 3] \,, \qquad (3.5), (3.6)$$

G, ν-Materialkonstanten. (3.6) gibt den einfachsten Mooney-Rivlin-Ansatz wieder [7]. (3.5), (3.6) in (3.2) bzw. (3.4) eingesetzt, liefern unter Berücksichtigung des Theorems von Hamilton-Cayley

$$\boldsymbol{B}^3 - \mathrm{I}_B \boldsymbol{B}^2 + \mathrm{II}_B \boldsymbol{B} - \mathrm{III}_B \boldsymbol{I} = 0 \,, \qquad \boldsymbol{T} = \frac{\varrho}{\varrho_0} G \left\{ \boldsymbol{B}^2 + \left[\left(1 + \frac{\nu}{1 - 2\nu} \right) (I_B - 3) + 2 - I_B \right] \boldsymbol{B} \right\} \qquad (3.7), (3.8)$$

bzw.

$$\boldsymbol{T} = -p\boldsymbol{I} + \tfrac{1}{2} G (\boldsymbol{B} - \boldsymbol{B}^{-1}) \,. \tag{3.9}$$

Bei kleinen Verzerrungen kann

$$B = I + 2E\,, \qquad |E_\varepsilon| \ll 1\,, \qquad \varepsilon = 1, 2, 3\,, \tag{3.10), (3.11}$$

vorausgesetzt werden. In Termen erster Ordnung in E lauten dann (3.8), (3.9)

$$\frac{\varrho}{\varrho_0} = [(1 + 2E_1)(1 + 2E_2)(1 + 2E_3)]^{-1/2}\,,$$

$$T = 2G\frac{\varrho}{\varrho_0}\left[E + \frac{\nu}{1-2\nu}I \operatorname{tr} E\right](I + 2E) \approx 2G\left[E + \frac{\nu}{1-2\nu}I \operatorname{tr} E\right] \tag{3.12}$$

bzw.

$$T \approx -pI + 2GE(I + E) \approx -pI + 2GE\,. \tag{3.13}$$

Die Gleichungen (3.12), (3.13) besitzen zwei wesentliche Eigenschaften. Sie sind invariant bezüglich starrer Bewegungen eines räumlichen Bezugssystems, und sie genügen (3.11). Es sei vermerkt, daß unter der Voraussetzung (3.11) jede beliebige positive Approximation $\Sigma, \tilde{\Sigma}$, die zu (3.12) bzw. (3.13) führt, anstelle von (3.5) bzw. (3.6) treten kann. Weiterhin äußert sich die willkürliche konstitutive Annahme eines anderen Verzerrungsmaßes und des im spezifischen Arbeitsdifferential zugeordneten Spannungsmaßes in unterschiedlichen Termen von nur zweiter und höherer Ordnung in E.

Mit der Spurbildung in (3.13) und $\varrho/\varrho_0 = 1$ ist auch zu sehen, daß

$$p - q = -\frac{4G}{3}(E_1E_2 + E_1E_3 + E_2E_3)$$

und deshalb

$$p - q \approx 0\,, \tag{3.14}$$

zur genäherten Erfüllung der Inkompressibilitätsbedingung durch (3.13) ausreicht.

Zur Berechnung von (1.19) werden die materiellen Zeitableitungen von (3.8), (3.9) gebildet:

$$\dot{T} = -G\frac{\varrho}{\varrho_0}\operatorname{tr} D\left\{B^2 + \left[\left(1 + \frac{\nu}{1-2\nu}\right)(\mathrm{I}_B - 3) + 2 - \mathrm{I}_B\right]B\right\} +$$
$$+ G\frac{\varrho}{\varrho_0}\left\{B\dot{B} + \dot{B}B + \frac{\nu}{1-2\nu}\dot{\mathrm{I}}_B B + \left[\left(1 + \frac{\nu}{1-2\nu}\right)(\mathrm{I}_B - 3) + 2 - \mathrm{I}_B\right]\dot{B}\right\}$$

bzw.

$$\dot{T} = -\dot{p}I + \tfrac{1}{2}G(\dot{B} + B^{-1}\dot{B}B^{-1})\,,$$

wo mit (1.5), (1.12), ..., (1.14), (3.10)

$$\dot{B} = (D + W)(I + 2E) + (I + 2E)(D + W^T) = 2D + 2(DE + ED) + 2(WE + EW^T)$$

ist und deshalb näherungsweise mit (3.12), (3.13)

$$\dot{T} \approx 2G\left[D + \frac{\nu}{1-2\nu}I \operatorname{tr} D\right] + WT + TW^T \tag{3.15}$$

bzw.

$$\dot{T} \approx -\dot{p}I + 2GD + WS + SW^T \tag{3.16}$$

gilt (E dabei zweckmäßig in Hauptachsendarstellung gedacht). (1.19) lautet also unter Berücksichtigung von (3.11), ..., (3.13) (d. h. Vernachlässigung von $T \operatorname{tr} D$ in (1.19))

$$\dot{T}_C \approx 2G\left(D + \frac{\nu}{1-2\nu}I \operatorname{tr} D\right) - TW - DT \tag{3.17}$$

bzw. mit (1.20)

$$\dot{T}_C \approx -\dot{q}I + 2GD\left(1 + \frac{q}{2G}\right) - TW - DS = -\dot{q}I + 2GD - SW + qL - DS\,. \tag{3.18}$$

(3.18) enthält einen Term $q/2G$, der in die Größenordnung von Eins kommen und damit das Verzweigungsgeschehen beeinflussen kann, ohne daß die Voraussetzung kleiner Verzerrungen verletzt wird. In der Gleichung des fortgesetzten Gleichgewichtes (2.1) entfällt der Term qL für $q =$ konst. wegen Erfüllung der Inkompressibilitätsbedingung (1.20). Die Wirkung dieses Terms in (2.2) hängt davon ab, wie sich das Oberflächenelement der Verzweigungslösung verformt. Wird dagegen die Änderung der wahren Spannung T auf dem Rande vorgegeben, so steht anstelle von (2.2) unter Berücksichtigung von (1.19), (1.20), (3.18)

$$(-T_{ln}L_{kn})\,n_k = (-\dot{q}\delta_{kl} + 2GD_{kl} - S_{kn}W_{nl} + qL_{kl} - D_{kn}S_{nl})\,n_k$$

(δ_{kl} — Kroneckersymbol), und diese Gleichung ist unabhängig von q. Mit (2.6), (1.20) folgt noch

$$L_{kl}L_{lk} = g_k e_l g_l e_k = 0\,,$$

so daß der Term qL^2 in (2.4), (2.5) nicht auftritt.

Die Gleichungen (3.15), (3.16) sind als Spezialfälle hypoelastischer Ansätze auffaßbar [7]. Ähnliche Spezialfälle werden, wie schon erwähnt, in der Literatur über Verzweigungsprobleme benutzt, z. B.

$$\dot{\boldsymbol{T}} = 2G\left(\boldsymbol{D} + \frac{\nu}{1-2\nu}\boldsymbol{I}\operatorname{tr}\boldsymbol{D}\right) + \boldsymbol{W}\boldsymbol{T} + \boldsymbol{T}\boldsymbol{W}^T + \alpha(\boldsymbol{D}\boldsymbol{T} + \boldsymbol{T}\boldsymbol{D}) - \beta\boldsymbol{T}\operatorname{tr}\boldsymbol{D} \tag{3.19}$$

und im inkompressiblen Fall

$$\dot{\boldsymbol{T}} = -\dot{q}\boldsymbol{I} + 2G\boldsymbol{D}\left(1 - \alpha\frac{q}{G}\right) + \boldsymbol{W}\boldsymbol{S} + \boldsymbol{S}\boldsymbol{W}^T + \alpha\left[\boldsymbol{D}\boldsymbol{S} + \boldsymbol{S}\boldsymbol{D} - \frac{2}{3}\boldsymbol{I}\operatorname{tr}(\boldsymbol{S}\boldsymbol{D})\right] \tag{3.20}$$

mit $\alpha = 0, \beta = 0$ [1, 3], $\alpha = 0, \beta = 1$ [2], $\alpha = 0, \beta = 1$ bzw. $\alpha = 1, \beta = 1$ [8]. Für

$$|T_\varepsilon/2G)| \ll 1\,, \qquad \varepsilon = 1, 2, 3\,, \tag{3.21}$$

gehen (3.19), (3.20) in (3.15), (3.16) über. Die erste Vernachlässigung ist auch in der bei der Herleitung von (3.15) aus (3.8) getroffenen Vernachlässigung der Terme

$$\boldsymbol{D}\boldsymbol{E} + \boldsymbol{E}\boldsymbol{D}\,, \qquad \boldsymbol{E}\operatorname{tr}\boldsymbol{D}\,, \qquad \boldsymbol{D}\operatorname{tr}\boldsymbol{E}\,, \qquad \boldsymbol{I}\operatorname{tr}\boldsymbol{E}\boldsymbol{D}\,, \qquad \boldsymbol{I}\operatorname{tr}\boldsymbol{E}\operatorname{tr}\boldsymbol{D}\,, \tag{3.22}$$

enthalten. Bei der Differentiation von (3.9) zu (3.16) mußte dagegen kein Term von (3.22) unterdrückt werden.

In den Verzweigungsbedingungen (2.4), (2.5) können unter der Voraussetzung (3.21) alle Terme (3.22) vernachlässigt werden, bzw. sie entfallen im inkompressiblen Fall wegen (1.20).

Bei Verletzung von (3.21) oder, falls $\alpha, \beta \gg 1$ (die speziellen Festlegungen von α, β in (3.19), (3.20) sind nicht begründet oder durch konventionelle Versuche mit finiter Deformation unter der Voraussetzung (3.11) motiviert worden), werden α, β konstitutive Parameter.

Der Einfluß der verschiedenen Materialannahmen auf eine spezielle Verzweigungslösung wird im folgenden Beispiel diskutiert.

4. Elastische Scherbandbildung in einer inkompressiblen Scheibe

Für einachsigen Zug $\sigma > 0$ in der Scheibenebene $x_3 =$ konst. und überlagerten hydrostatischen Druck $-k\sigma$, $k > 0$, sind die Hauptspannungen in kartesischen Koordinaten bestimmt durch

$$T_{kl} = \begin{bmatrix} 1-k & & \\ & -k & \\ & & -k \end{bmatrix} \sigma\,. \tag{4.1}$$

Das Scherband besitze z. B. [1] die Orientierung

$$e_k = (e_1\ 0\ e_3)\,. \tag{4.2}$$

Die Inkompressibilitätsbedingung lautet mit (2.6), (4.2)

$$g_1 e_1 + g_3 e_3 = 0\,. \tag{4.3}$$

Diese drei Gleichungen und die Elastizitätsgesetze (3.18), (3.20) in (1.19) ergeben mit den Verzweigungsbedingungen (2.5), (2.6) sowie der Abkürzung (4.4)

$$e_1/e_3 = e\,, \qquad (2G + \sigma)\,e^4 + 4Ge^2 + 2G - \sigma = 0\,, \qquad |S_\varepsilon/G| \ll 1\,, \tag{4.4), (4.5}$$

$$(2G + \sigma)\,e^4 + 4Ge^2 + 2G - \sigma = 0\,, \qquad \alpha = 0\,, \tag{4.6}$$

$$(2G - 2k\sigma + 2\sigma)\,e^4 + (4G - 4k\sigma + 2\sigma)\,e^2 + 2G - 2k\sigma = 0\,, \qquad \alpha = 1\,. \tag{4.7}$$

Für (4.5), wo im übrigen k erwartungsgemäß nicht mehr auftritt, muß wegen (3.11)

$$\sigma/G \ll 1 \tag{4.8}$$

sein. Deshalb liefert (4.5) keine reelle Lösung e. Die Ab- oder Anwesenheit von k in (4.6), (4.7) wurde von α gesteuert. Wird die Voraussetzung (4.8) fallengelassen, ergibt (4.6) die Lösung

$$\sigma = 2G\,, \qquad e = 0\,. \tag{4.9}$$

Diese gilt wie in [1] ebenso für reinen einachsigen Zug. (4.7) besitzt reelle Lösungen, ohne daß $\sigma \ll G$ verletzt wird, wenn $k\sigma$ etwa gleich G ist. Der zweite Term auf der rechten Seite von (3.20) verringert dann seine Größenordnung, und die Lösungsmannigfaltigkeit von (4.7) wird auch gegenüber Modulschwankungen und Vernachlässigungen von Spannungs- gegenüber Modultermen empfindlich. Z. B. ergibt

$$k\sigma = G + \tfrac{3}{4}\,\sigma\,, \tag{4.10}$$

in (4.7)

$$e^4 - 2e^2 - 3 = 0 \tag{4.11}$$

bzw.

$$e^2 = 3 \tag{4.12}$$

unabhängig von Angaben über σ. Der Parameter α übt deshalb unter diesen Bedingungen einen konstitutiven Einfluß auf das Lösungsverhalten der Verzweigungsbedingung aus.

Zur nachträglichen Kontrolle der Grunddeformation im obigen Beispiel wird (4.1) in (3.20) eingesetzt. Das Ergebnis ist

$$\frac{\dot{\sigma}}{3G} = D_{11}\left[1 + \alpha \frac{\sigma}{G}\left(\frac{2}{3} - k\right)\right] \tag{4.13}$$

bzw. nach Integration

$$E_{\log} = \frac{\ln\left|1 + \alpha \frac{\sigma}{G}\left(\frac{2}{3} - k\right)\right|}{3\alpha\left(\frac{2}{3} - k\right)} \approx \frac{\sigma}{3G}\left[1 - \frac{\alpha\sigma}{2G}\left(\frac{2}{3} - k\right)\right], \tag{4.14}$$

wo $E_{\log} = \ln (l/L)$ die logarithmische Dehnung bezeichnet und die Näherung nur für kleine Argumente

$$\left|\frac{\alpha\sigma}{G}\left(\frac{2}{3} - k\right)\right| \ll 1$$

gilt. Gleichung (4.14) zeigt, wie der hydrostatische Druck die Grunddeformation in Abhängigkeit von α beeinflußt. Bei $\alpha = 0$ verschwindet diese Wirkung. Zum Beispiel liefern $\alpha = 1$ und (4.10) in (4.14)

$$\frac{l}{L} = \left(12\frac{G}{\sigma}\right)^{\frac{1}{3\frac{G}{\sigma} + \frac{1}{4}}} > 1$$

mit

$$\lim_{G/\sigma \to \infty} (l/L) = 1 .$$

5. Schlußfolgerungen

Die oben besprochenen Materialtypen gestatten den nachstehenden zusammenfassenden Vergleich für ihre Verwendung in den klassischen Verzweigungsbedingungen.

a) Kompressibilität

	$\lvert\alpha\rvert, \lvert\beta\rvert = O(1)$		$\lvert\alpha\rvert, \lvert\beta\rvert \gg 1$
	$\lvert T_\varepsilon\rvert \ll G$	$\lvert T_\varepsilon/G\rvert = O(1)$	$\lvert T_\varepsilon\rvert \ll G$
Hyperelastizität (3.15)	Äquivalenz	verboten	—
Hypoelastizität (3.19)		α, β konstitutiv	α, β konstitutiv

b) Inkompressibilität

	$\lvert\alpha\rvert = O(1)$			$\lvert\alpha\rvert \gg 1$
	$\lvert T_\varepsilon\rvert \ll G$	$\lvert S_\varepsilon\rvert \ll G$	$\lvert S_\varepsilon/G\rvert = O(1)$	$\lvert S_\varepsilon\rvert \ll G$
Hyperelastizität (3.16)	Äquivalenz	q bedeutsam je nach Randbedingungen	verboten	—
Hypoelastizität (3.20)		q bedeutsam, α konstitutiv	α konstitutiv	α konstitutiv

$O(1)$ bezeichnet die Größenordnung von 1.

Es sei noch darauf hingewiesen, daß die Angabe von Verzweigungslösungen, die von den konstitutiven Parametern α, β abhängen, entweder diese Parameter selbst oder deren experimentelle Begründung enthalten muß.

Literatur

1 NEMAT-NASSER, S., Finite deformation plasticity and plastic instability, Twenty-fifth Conference of Army Mathematicians Baltimore, 1979.
2 NEEDLEMAN, A.; TVERGAARD, V., Necking of biaxially stretched elastic-plastic circular plates, J. Mech. Phys. Sol. **25**, 159—183 (1977).
3 CHENG, S. Y.; S. T. ARIARATNAM; R. N. DUBEY, Axisymmetric bifurcation in an elastic-plastic cylinder under axial load and lateral hydrostatic pressure. Quart. Appl. Math. **29**, 41—51 (1971).
4 RICE, J. R., The localization of plastic deformation, in: Proc. 14th IUTAM Congress on Theoretical and Applied Mechanics (KOITER, W. T., Hrsg.). Amsterdam 1976, 207—220.
5 HILL, R., On uniqueness and stability in the theory of finite elastic strain, J. Mech. Phys. Sol. **5**, 229—241 (1957).
6 TRUESDELL, C.; R. A. TOUPIN, The classical field theories in: Handbuch der Physik, Bd. III/1 (FLÜGGE, S. Hrsg.). Berlin 1960,
7 TRUESDELL, C.; W. NOLL, The non-linear field theories of mechanics, in: Handbuch der Physik, Bd. III/3 (FLÜGGE, S., Hrsg)., Berlin 1965.

8 HUTCHINSON, J. W.; J. P. MILES, Bifurcation analysis of the onset of necking in an elastic-plastic cylinder under uniaxial tension, J. Mech. Phys. Sol. 22, 61–71 (1974).
9 SPENCER, A. J. M.; J. E. FERRIER, Some solutions for a class of plastic-elastic solids, in: Foundations of Plasticity (SAWCZUK, A., Hrsg.), Leyden 1973.
10 HILL, R., On the classical constitutive relations for elastic-plastic solids, in: Recent Progress in Applied Mechanics (BROBERG B., u. a. Hrsg.). Stockholm 1967, 241–249.

Eingereicht am 10. 7. 1981

Anschrift: Dr.-Ing. H. BALKE, Zentralinstitut für Festkörperphysik und Werkstofforschung der Akademie der Wissenschaften der DDR, 8027 Dresden, Postfach, DDR

ZAMM 62, 249 – 255 (1982)

T. K. DE

On the Phase Boundary Motion in the Earth Due to Pressure and Temperature Excitations at the Earth's Surface

Assuming that a change of pressure and temperature brings out phase transitions inside the Earth, the motion of the phase boundary inside the Earth due to pressure and temperature excitations at the Earth's surface has been determined under linear approximation. It is shown in this paper that the solution of this Stefan-like problem consists not only of the phase boundary motion due to pressure and temperature excitation at the Earth's surface but also of the phase boundary motion due to different thermal properties at the two phases. Moreover, the phase boundary motion due to temperature excitation is found to be dependent on the corresponding pressure excitation.

Unter der Annahme, daß eine Druck- und Temperaturänderung Phasenübergänge im Erdinneren erzeugt, wurde die durch Druck- und Temperaturerregung an der Erdoberfläche hervorgerufene Bewegung der Phasenfläche im Erdinneren durch lineare Approximation bestimmt. In dieser Arbeit wird gezeigt, daß die Lösung dieses Problems vom Stefanschen Typ nicht nur aus der durch Druck- und Temperaturerregung an der Erdoberfläche hervorgerufenen Phasenflächenbewegung, sondern auch aus der von den unterschiedlichen thermischen Eigenschaften in den beiden Phasen erzeugten Phasenflächenbewegung besteht. Darüberhinaus ergab sich, daß die durch Temperaturerregung hervorgerufene Phasenflächenbewegung von der entsprechenden Druckerregung abhängt.

При условии, что изменения давления и температуры вызывают фазовые переходы внутри земли было с помощью линейной аппроксимации определено движение фазовых поверхностей внутри земли, вызываемое возбуждением температуры и давления на поверхности земли. В этой работе показоно, что решение этой задачи типа Стефана не только состоит из движения фазовых поверхностей, вызываемого возбуждением температуры и давления на поверхности земли, но и из движения фазовых поверхностей, полученного неодинаковыми термическими свойствами в двух фазах. Кроме того найдено, что движение фазовых поверхностей, вызываемое возбуждением температуры зависит от соответственного возбуждения давления.

Introduction

Geological observations indicate that vertical crustal movements take place in different regions of the Earth. Although the driving mechanism for vertical crustal motion can not be explained in all cases, in some cases the driving force is definitely gravity. However, the cause of this vertical crustal motion in many cases can be explained, atleast partially, as a response of the phase boundaries within the Earth due to changes in pressure and temperature.

The determination of the phase-boundary motion due to pressure and temperature excitations is similar to STEFAN's problem (CARLAN and JAEGER, 1959). Special solutions to this STEFAN-like problem by numerical methods have been obtained by MCDONALD and NESS (1960), VAN DE LINDT (1967), JOYNER (1967). Analytical solutions under different approximations of this problem have been obtained by O'CONNELL and WASSERBERG (1967, 1972), GJEVIK (1972, 1973) and MARESCHAL and GANGI (1977).

The main purpose of this paper is to show that (i) the solution of the problem under linear approximations can be obtained by a much simpler approach than that of the previous authors (ii) the solution of the problem consists not only of the phase boundary motion due to pressure and temperature excitations as shown by MARESCHAL and GANGI but also of the phase boundary motion due to different thermal properties of the two phases (iii) the phase boundary motion due to temperature excitation, which has been shown by MARESCHAL and GANGI to be independent of the corresponding pressure excitation, is found to be dependent on the corresponding pressure excitation.

Statement of the problem

We assume that two phases exist inside the Earth and it is expected that a change of pressure and temperature may bring out phase transitions.

The phase boundary motion has been considered here in a system of co-ordinates fixed to the Earth's surface. We take $x = 0$ as the Earth's surface and $x = x_m(t)$ as the phase boundary.

The pressure P_m and the temperature T_m for which two phases are in equilibrium are related by the CLAUSIUS-CLAPEYRON equation which we assume to be linear. Therefore we have at the interface

$$T_m = T_c + \nu P_m , \tag{1}$$

where T_c is the transition temperature at the Earth's surface and ν is the inverse slope of the CLAUSIUS-CLAPEYRON equation. If $P_0(t)$ is the pressure applied on the Earth's surface, we have

$$T_m = T_c + \nu P_0(t) + \nu g \varrho_1 x_m(t) , \tag{2}$$

where ϱ_1 is the constant density of the phase between the surface and the transition point.

In order to hold the law of heat conservation during the phase boundary motion, we must have [CARSLAW and JAEGER, 1959]

$$L\varrho_1 \frac{\mathrm{d}x_m}{\mathrm{d}t} = k_2 \frac{\partial T_2}{\partial x}\bigg|_{x_m} - k_1 \frac{\partial T_1}{\partial x}\bigg|_{x_m} , \tag{3}$$

where L is the latent heat of transformation, T_1 and T_2 are the temperatures in each phases.

The motion of the phase boundary will depend on how this extra heat is carried away and this is determined from the solution of the heat equation:

$$\frac{\partial T_1}{\partial t} = H_1 \frac{\partial^2 T_1}{\partial x^2}, \qquad x < x_m(t), \quad \text{i.e.} \quad s < 0, \tag{4a}$$

$$\frac{\partial T_2}{\partial t} + v \frac{\partial T_2}{\partial x} = H_2 \frac{\partial^2 T_2}{\partial x^2}, \qquad x > x_m(t), \quad \text{i.e.} \quad s > 0, \tag{4b}$$

where $s = x - x_m(t)$ and H_1 and H_2 are the thermal diffusivity of two planes and v is the vertical component of the velocity with which the deepest phase moves.

We assume that no heat sources are present and the system is initially in thermal equilibrium. The initial temperature distribution, which we assume to be a linear function of depth, can be written as

$$T_1(x, t = 0) = \beta_1 x, \tag{5a}$$

$$T_2(x, t = 0) = \beta_1 x_0 + \beta_2 (x - x_0), \tag{5b}$$

where the initial phase boundary depth x_0 is given by

$$\beta_1 x_0 = T_c + \gamma g \varrho_1 x_0 . \tag{6}$$

The boundary conditions are

$$\text{(i)} \quad T_1(x = 0, t) = T_0(t), \tag{7a}$$

$$\text{(ii)} \quad \lim_{x \to \infty} \frac{\partial T_2}{\partial x} = \beta_2, \tag{7b}$$

$$\text{(iii)} \quad T_1|_{x_m} = T_2|_{x_m} = T_m, \tag{7c}$$

i.e. at the interface between two phases the temperature is continuous and is equal to the transition temperature given by (2).

(iv) At the interface the heat flow is discontinuous and obeys STEFAN's boundary condition given by (3) (7d)

Mathematical solution

Assuming that the phase-boundary motion is very small, we can neglect the second order convection term $v \frac{\partial T_2}{\partial x}$. Therefore, we assume the heat flow to be linear and thus the system of equations in (4a-b) has been linearised.

We introduce the LAPLACE transform defined by

$$\overline{T} = \int_0^\infty T(x, t)\, e^{-pt}\, dt, \qquad \operatorname{Re} p > 0 . \tag{8}$$

The relation

$$s = x - x_m(t)$$

can now be written as

$$\overline{s} = \frac{x}{p} - \overline{x}_m . \tag{9}$$

Utilising the initial conditions (5a-b) and (7b), the solution of the transformed equations (4a-b) can be written as

$$\overline{T}_1 = A\, e^{m_1 x} + B\, e^{-m_1 x} + \frac{\beta_1}{p} x, \qquad \frac{x}{p} < \overline{x}_m, \quad \text{i.e.} \quad \overline{s} < 0, \tag{10a}$$

$$\overline{T}_2 = D\, e^{-m_2 x} - \frac{(\beta_2 - \beta_1)}{p} x_0 + \frac{\beta_2}{p} x, \qquad \frac{x}{p} > \overline{x}_m, \quad \text{i.e.} \quad \overline{s} > 0, \tag{10b}$$

where

$$m_1 = \sqrt{p/H_1}, \qquad m_2 = \sqrt{p/H_2} . \tag{11}$$

Taking the LAPLACE transform of the boundary conditions (7a), (7c) [i.e. $\overline{T}_1|_{(x/p)=\overline{x}_m} = \overline{T}_2|_{(x/p)=\overline{x}_m}$] and (3) and utilising these boundary conditions, we get

$$A + B = \overline{T}_0(p), \tag{12a}$$

$$A\, e^{m_1 p \overline{x}_m} + B\, e^{-m_1 p \overline{x}_m} - (\beta_2 - \beta_1)\left(\overline{x}_m - \frac{x_0}{p}\right) = D\, e^{-m_2 p \overline{x}_m}, \tag{12b}$$

$$\frac{k_1 m_1}{k_2 m_2}\left[A\, e^{m_1 p \overline{x}_m} - B e^{-m_1 p \overline{x}_m}\right] - \frac{(k_2 \beta_2 - k_1 \beta_1)}{k_2 m_2 p} + \frac{L \varrho_1 p}{k_2 m_2}\left(\overline{x}_m - \frac{x_0}{p}\right) = - D\, e^{-m_2 p \overline{x}_m} . \tag{12c}$$

Adding (12b) and (12c) and solving this resultant equation together with (12a) for A and B, we have

$$\frac{A}{\overline{X} - b\overline{T}_0\, e^{-m_1 p \bar{x}_m}} = \frac{B}{-\overline{X} + a\overline{T}_0\, e^{m_1 p \bar{x}_m}} = \frac{1}{a\, e^{m_1 p \bar{x}_m} - b\, e^{-m_1 p \bar{x}_m}} \tag{13}$$

where

$$\overline{X} = \frac{k_2\beta_2 - k_1\beta_1}{k_2 p^{3/2}} + (\beta_2 - \beta_1)\left(\bar{x}_m - \frac{x_0}{p}\right) - \frac{L\varrho_1\sqrt{H_2}}{k_2}\cdot p^{1/2}\cdot\left(\bar{x}_m - \frac{x_0}{p}\right),$$

$$a = 1 + \frac{k_1 m_1}{k_2 m_2} = 1 + \frac{k_1}{k_2}\cdot\sqrt{\frac{H_2}{H_1}}, \qquad b = 1 - \frac{k_1 m_1}{k_2 m_2} = 1 - \frac{k_1}{k_2}\cdot\sqrt{\frac{H_2}{H_1}}. \tag{14}$$

A and B being known, we can find out D either from (12b) or from (12c).

The boundary condition (7c), i.e.

$$\overline{T}_m|_{(x/p)=\bar{x}_m} = \overline{T}_1|_{(x/p)=\bar{x}_m} = \overline{T}_2|_{(x/p)=\bar{x}_m},$$

gives

$$\frac{\overline{X}(e^{m_1 p\bar{x}_m} - e^{-m_1 p\bar{x}_m}) + (a-b)\,\overline{T}_0}{a\, e^{m_1 p\bar{x}_m} - b\, e^{-m_1 p\bar{x}_m}} = \nu\overline{P}_0(p) + (\nu g\varrho_1 - \beta_1)\left(\bar{x}_m - \frac{x_0}{p}\right). \tag{15}$$

Let us put

$$\bar{z}_m = \bar{x}_m - \frac{x_0}{p}. \tag{16}$$

Therefore

$$e^{\pm m_1 p\bar{x}_m} = e^{\pm m_1 x_0}\cdot e^{\pm m_1 p\bar{z}_m} = e^{\pm m_1 x_0}\,[1 \pm p m_1\bar{z}_m \pm \cdots].$$

Since $\bar{z}_m$ is small, neglecting second and higher order terms we can write

$$e^{\pm m_1 p\bar{x}_m} \simeq e^{\pm m_1 x_0}\,[1 \pm p m_1 \bar{z}_m]. \tag{17}$$

Utilising (17) and simplifying (15), we get ultimately

$$\bar{z}_m = \frac{1}{a(\nu g\varrho_1 - \beta_1)\, R(p)}\Bigg[-(a - b\, e^{-2\sqrt{\tau_0 p}})\,\nu\overline{P}_0 + (a-b)\,\overline{T}_0\, e^{-\sqrt{\tau_0 p}} + \frac{(k_2\beta_2 - k_1\beta_1)\sqrt{H_2}}{k_2}\cdot\frac{1}{p^{3/2}}\cdot(1 - e^{-2\sqrt{\tau_0 p}})\Bigg], \tag{18}$$

where

$$R(p) = \sqrt{\tau_1 p} + \frac{\nu p^{3/2}\overline{P}_0}{(\nu g\varrho_1 - \beta_1)\sqrt{H_1}} + 1 - \varepsilon_1 - \left[\sqrt{\tau_1 p} - \frac{b\nu p^{3/2}\overline{P}_0}{a(\nu g\varrho_1 - \beta_1)\sqrt{H_1}} + \varepsilon_2\right] e^{-2\sqrt{\tau_0 p}},$$

$$\tau_0 = \frac{x_0^2}{H_1}, \qquad \tau_1 = \left[\frac{L\varrho_1\sqrt{H_2}}{a k_2(\nu g\varrho_1 - \beta_1)}\right]^2, \tag{19}$$

$$\varepsilon_1 = \frac{(k_2\beta_2 - k_1\beta_1)}{a k_2(\nu g\varrho_1 - \beta_1)}\cdot\sqrt{\frac{H_2}{H_1}} - \frac{(\beta_2 - \beta_1)}{a(\nu g\varrho_1 - \beta_1)}, \qquad \varepsilon_2 = \varepsilon_1 + \frac{b}{a}.$$

It is obvious from (14) and (19) that ε_1 and ε_2 are very small order terms which can be neglected.

The phase boundary motion can now be simply obtained from (18) by taking the inverse of the LAPLACE transform. The first and the second term on the right hand side of (18) correspond to phase boundary motion due to pressure and temperature excitations. The third term on the right hand side of (18) represents the phase boundary motion due to different thermal properties at the two phases which is absent in MARESCHAL's and GANGI's solution. However, the phase boundary motion due to different thermal properties at the two phases is very small. Moreover, the phase boundary motions due to temperature excitation and different thermal properties at the two phases are always dependent on the corresponding pressure excitation since $R(p)$ contains the term $\overline{P}_0(p)$.

If we assume that the thermal properties at the two phases are the same, i.e.

$$k_1 = k_2 = k\,, \qquad H_1 = H_2 = H\,, \qquad \beta_1 = \beta_2 = \beta\,, \tag{20}$$

and if we take $e^{\pm m_1 p\bar{x}_m} \simeq e^{\pm m_1 x_0}$, then $\bar{z}_m$ is given by

$$\bar{z}_m = \frac{-\nu\overline{P}_0 + \overline{T}_0\, e^{-\sqrt{\tau_0 p}}}{(\nu g\varrho_1 - \beta_1)\,[\sqrt{\tau_1 p} + 1 - \sqrt{\tau_1 p}\, e^{-2\sqrt{\tau_0 p}}]}$$

which is in complete agreement with the result of MARESCHAL and GANGI (1977).

Phase boundary motion due to pressure excitation

We take

$$P_0(t) = P_0 H\left(\frac{t}{\tau}\right)$$

where

$$\sqrt{\tau} = \sqrt{\tau_1} + \sqrt{\tau_2}\,, \qquad \sqrt{\tau_2} = \frac{\nu P_0}{(\nu g \varrho_1 - \beta_1)\sqrt{H_1}}\,, \tag{21}$$

and $H\left(\frac{t}{\tau}\right)$ is the Heaviside function.

Neglecting the small order terms ε_1 and ε_2, the phase boundary motion for the given pressure excitation can be written from (18) as

$$\frac{z_m(t)}{z_0} = \frac{1}{2\pi i} \int\limits_{\sigma - i\infty}^{\sigma + i\infty} \frac{\left[1 - \frac{b}{a} e^{-2\sqrt{\tau_0 p}}\right]}{\sqrt{\tau p} + 1 - \sqrt{\tau_3 p} \cdot e^{-2\sqrt{\tau_0 p}}} e^{pt} \cdot dp\,, \tag{22}$$

where

$$\sqrt{\tau_3} = \sqrt{\tau_1} - \frac{b}{a}\sqrt{\tau_2}\,, \qquad z_0 = -\nu P_0/(\nu g \varrho_1 - \beta_1)\,. \tag{23}$$

The integral (22) can be replaced by an integral along the branch out of $\sqrt{p}$ as follows

$$\frac{z_m(t)}{z_0} = \frac{1}{\pi} \int\limits_0^\infty \frac{1}{F(r)} \left[\sqrt{\tau r} - \sqrt{\tau_3 r} \cos 2\sqrt{\tau_0 r}\right] \frac{1}{r} \cdot e^{-rt} \cdot dr - \frac{1}{\pi} \cdot \frac{b}{a} \int\limits_0^\infty \frac{1}{F(r)} \left[(1 - \sqrt{\tau_3 r} \sin 2\sqrt{\tau_0 r}) \sin 2\sqrt{\tau_0 r} + \right.$$
$$\left. + (\sqrt{\tau r} - \sqrt{\tau_3 r} \cos 2\sqrt{\tau_0 r}) \cos 2\sqrt{\tau_0 r}\right] \frac{1}{r} \cdot e^{-rt}\,, \tag{24}$$

where

$$F(r) = 1 + r(\tau + \tau_3) - 2r\sqrt{\tau \tau_3} \cos 2\sqrt{\tau_0 r} - 2\sqrt{\tau_3 r} \sin 2\sqrt{\tau_0 r}\,, \tag{25}$$

which can be evaluated by numerical integration.

Instead of integrating (25) numerically, we can expand the integral of (22) into a series as

$$\frac{z_m(t)}{z_0} = \frac{1}{2\pi i} \int\limits_{\sigma - i\infty}^{\sigma + i\infty} \frac{1 - \frac{b}{a} \cdot e^{-2\sqrt{\tau_0 p}}}{\sqrt{\tau p} + 1} \sum_{n=0}^{\infty} \left(\frac{\sqrt{\tau_3 p}}{1 + \sqrt{\tau_1 p}}\right)^n \cdot e^{-2n\sqrt{\tau_0 p}} \cdot e^{pt} \cdot dp\,, \tag{26}$$

which can now be integrated term by term. Retaining the first two terms, which is sufficient for practical interest, we get

$$\frac{z_m(t)}{z_0} = 1 - e^{\sqrt{t/\tau}} \operatorname{Erfc}\left(\sqrt{\frac{t}{\tau}}\right) + 2\sqrt{\frac{\tau_3}{\tau}}\left[\sqrt{\frac{t}{\pi\tau}}\, e^{-\tau_0/t} - \left(\frac{t}{\tau} + \sqrt{\frac{\tau_0}{\tau}}\right) e^{2\sqrt{\tau_0/\tau} + t/\tau} \cdot \operatorname{Erfc}\left(\sqrt{\frac{\tau_0}{t}} + \sqrt{\frac{t}{\tau}}\right)\right] -$$
$$- \frac{b}{a}\left[\operatorname{Erfc}\left(\sqrt{\frac{\tau_0}{t}}\right) - e^{2\sqrt{\tau_0/\tau} + t/\tau} \cdot \operatorname{Erfc}\left(\sqrt{\frac{\tau_0}{t}} + \sqrt{\frac{t}{\tau}}\right) - 2\sqrt{\frac{\tau_3}{\tau}}\left\{\sqrt{\frac{t}{\pi\tau}}\, e^{-4\tau_0/t} - \right.\right.$$
$$\left.\left. - \left(\frac{t}{\tau} + 2\sqrt{\frac{\tau_0}{\tau}}\right) e^{4\sqrt{\tau_0/\tau} + t/\tau} \cdot \operatorname{Erfc}\left(2\sqrt{\frac{\tau_0}{t}} + \sqrt{\frac{t}{\tau}}\right)\right\}\right], \tag{27}$$

where Erfc is the complementary error function defined by

$$\operatorname{Erfc}(x) = 1 - \operatorname{Erfc}(x) = 1 - \frac{2}{\sqrt{\pi}} \int\limits_0^x e^{-t^2} \cdot dt.$$

The phase boundary motion due to pressure excitation is shown in the figure 1 under the assumption that the thermal properties at the two phase are the same [i.e. under the assumptions (20)] so that $b = 0$, $\tau_3 = \tau_1$.

Phase boundary motion due to temperature excitation corresponding to pressure excitation $P_0 H(\tau/\tau)$

Case I

$$T_0(t) = T_0 \cdot \delta(t/\tau)\,,$$

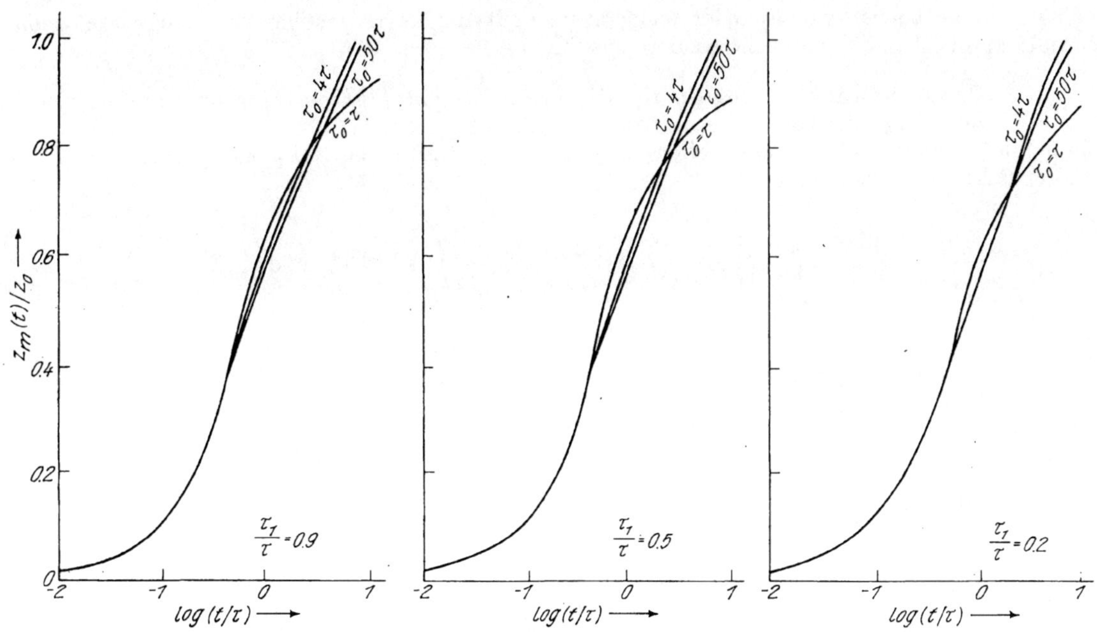

Fig. 1. Phase boundary motion due to a Heaviside function pressure excitation

where $\delta(t/\tau)$ is the DIRAC delta function. As in the previous case, the phase boundary motion is given by

$$\frac{z_m(t)}{z_0} = -\frac{\tau\left(1-\frac{b}{a}\right)}{\pi}\int\limits_0^\infty \frac{1}{F(r)}\left[(1-\sqrt{\tau_3 r}\sin 2\sqrt{\tau_0 r})\sin 2\sqrt{\tau_0 r} + (\sqrt{\tau r}-\sqrt{\tau_3 r}\cos 2\sqrt{\tau_0 r})\cos 2\sqrt{\tau_0 r}\right]\cdot e^{-rt}\,dr \tag{28}$$

where

$$z_0 = \frac{T_0}{\nu g \varrho_1 - \beta_1} \tag{29}$$

and $F(r)$ is given by (26).

However, the phase boundary motion can also be written as

$$\frac{z_m(t)}{z_0} = \frac{\tau\left(1-\frac{b}{a}\right)}{2\pi i}\int\limits_{\sigma-i\infty}^{\sigma+i\infty} \frac{e^{-\sqrt{\tau_0 p}}}{\sqrt{\tau p}+1}\sum_{n=0}^{\infty}\left(\frac{\sqrt{\tau_3 p}}{\sqrt{\tau p}+1}\right)^n \cdot e^{-2n\sqrt{\tau_0 p}}\, e^{pt}\,dp\,, \tag{30}$$

which can be integrated term by term.

Retaining the first two terms of the series, we get

$$\frac{z_m(t)}{z_0} = \left(1-\frac{b}{a}\right)\left[\sqrt{\frac{\tau}{\pi t}}\cdot e^{-\tau_0/4t} - e^{\sqrt{\tau_0/\tau}+t/\tau}\cdot \mathrm{Erfc}\left(\frac{1}{2}\sqrt{\frac{\tau_0}{t}}+\sqrt{\frac{t}{\tau}}\right)\right] + \left(1-\frac{b}{a}\right)\sqrt{\frac{\tau_3}{\tau}}\left[\left(\sqrt{\frac{\tau}{\pi t}}+2\sqrt{\frac{t}{\pi\tau}}\right)e^{-q\tau_0/4t} - \left(2+2\frac{t}{\tau}+3\sqrt{\frac{\tau_0}{t}}\right)e^{3\sqrt{\tau_0/\tau}+t/\tau}\ \mathrm{Erfc}\left(\frac{3}{2}\sqrt{\frac{\tau_0}{t}}+\frac{t}{\tau}\right)\right]. \tag{31}$$

Case II

$T_0(t) = T_0 \cdot H(t/\tau)$.

The phase boundary motion is given by

$$\frac{z_m(t)}{z_0} = \frac{\tau\left(1-\frac{b}{a}\right)}{2\pi i}\int\limits_{\sigma-i\infty}^{\sigma+i\infty} \frac{e^{-\sqrt{\tau_0 p}}}{p(\sqrt{\tau p}+1)}\sum_{n=0}^{\infty}\left(\frac{\sqrt{\tau_3 p}}{\sqrt{\tau p}+1}\right)^n \cdot e^{-2n\sqrt{\tau_0 p}}\cdot e^{pt}\cdot dp\,. \tag{32}$$

Retaining the first two terms of the series, we obtain

$$\frac{z_m(t)}{z_0} = \left(1-\frac{b}{a}\right)\left[\mathrm{Erfc}\left(\frac{1}{2}\sqrt{\frac{\tau_0}{t}}\right) - e^{\sqrt{\tau_0/\tau}+t/\tau}\cdot \mathrm{Erfc}\left(\frac{1}{2}\sqrt{\frac{\tau_0}{t}}+\sqrt{\frac{t}{\tau}}\right)\right] + \left(1-\frac{b}{a}\right)\sqrt{\frac{\tau_3}{\tau}}\left[2\sqrt{\frac{t}{\pi\tau}}e^{-q\tau_0/4t} - \left(2\frac{t}{\tau}+3\sqrt{\frac{\tau_0}{\tau}}\right)e^{3\sqrt{\tau_0/\tau}+t/\tau}\cdot \mathrm{Erfc}\left(\frac{3}{2}\sqrt{\frac{\tau_0}{t}}+\frac{t}{\tau}\right)\right] \tag{33}$$

The phase boundary motion for temperature excitation corresponding to pressure excitation $P_0H(t/\tau)$ is shown in figures 2 and 3, under the assumptions (20), i.e. $b = 0$, $\tau_3 = \tau_1$.

Phase boundary motion for different thermal properties at the two phases corresponding to pressure excitation $P_0H(t/\tau)$

The phase boundary motion due to different thermal properties of the two phases corresponding to pressure excitation $P_0H(t/\tau)$ is given by

$$\frac{z_m(t)}{z_0} = \frac{1}{2\pi i\sqrt{\tau}} \int\limits_{\sigma-i\infty}^{\sigma+i\infty} \frac{\left(1 - e^{-2\sqrt{\tau_0 p}}\right)}{\left(\sqrt{\tau p}+1\right) p^{3/2}} \cdot \sum_{n=0}^{\infty} \left(\frac{\sqrt{\tau_3 p}}{\sqrt{\tau p}+1}\right)^n \cdot e^{-2n\sqrt{\tau_0 p}} \cdot e^{pt}\, dp\,,$$

where

$$z_0 = \frac{(k_2\beta_2 - k_1\beta_1)\sqrt{H_2\tau}}{k_2} \cdot \frac{1}{a(\nu g\varrho_1 - \beta_1)}\,.$$

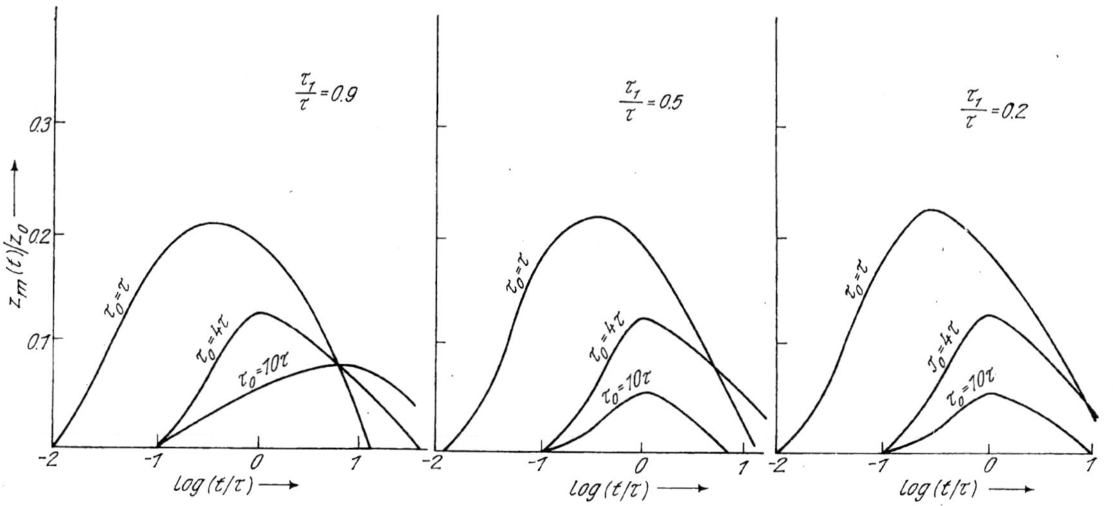

Fig. 2. Phase boundary motion due to a Dirac delta function temperature excitation corresponding to a Heaviside function pressure excitation

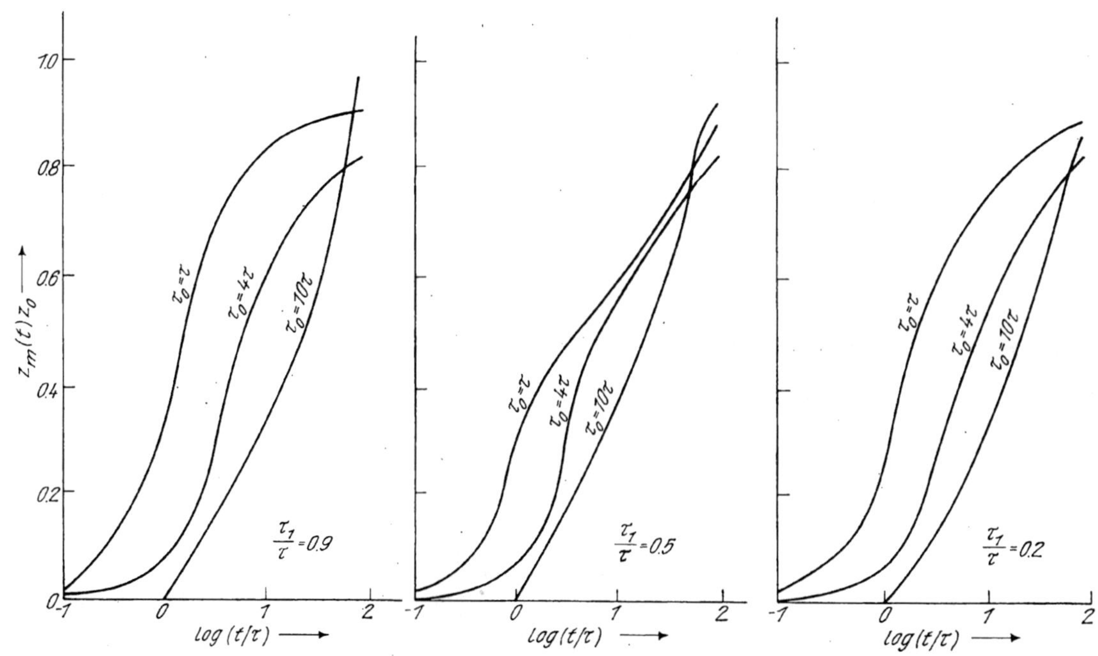

Fig. 3. Phase boundary motion due to a Heaviside function temperature excitation corresponding to a Heaviside function pressure excitation

Retaining the first two terms of the series, we have

$$\begin{aligned}\frac{z_m(t)}{z_0} = & -1 + 2\sqrt{\frac{t}{\pi\tau}} \cdot \left(1 - e^{-\tau_0/t}\right) + e^{t/\tau} \cdot \operatorname{Erfc}\left(\sqrt{\frac{t}{\tau}}\right) - \left(1 + 2\sqrt{\frac{\tau_0}{\tau}}\right) \operatorname{Erfc}\left(\sqrt{\frac{\tau_0}{t}}\right) + \\ & + e^{2\sqrt{\tau_0/\tau} + \sqrt{t/\tau}} \cdot \operatorname{Erfc}\left(\sqrt{\frac{\tau_0}{t}} + \sqrt{\frac{t}{\tau}}\right) + \\ & + \sqrt{\frac{\tau_3}{\tau}} \left[\operatorname{Erfc}\left(\frac{\tau_0}{t}\right) - 2\sqrt{\frac{t}{\pi\tau}}\, e^{-\tau_0/t} + \left(2\frac{t}{\tau} + 2\sqrt{\frac{\tau_0}{\tau}} - 1\right) e^{2\sqrt{\tau_0/\tau} + t/\tau} \cdot \operatorname{Erfc}\left(\sqrt{\frac{\tau_0}{t}} + \sqrt{\frac{t}{\tau}}\right) - \right. \\ & \left. - \operatorname{Erfc}\left(2\sqrt{\frac{\tau_0}{t}}\right) + 2\sqrt{\frac{t}{\pi\tau}} \cdot e^{-4\tau_0/t} - \left(2\frac{t}{\tau} + 4\sqrt{\frac{\tau_0}{\tau}} - 1\right) \cdot e^{4\sqrt{\tau_0/\tau} + \sqrt{t/\tau}} \cdot \operatorname{Erfc}\left(2\sqrt{\frac{\tau_0}{t}} + \sqrt{\frac{t}{\tau}}\right)\right].\end{aligned} \tag{34}$$

However the phase boundary motion due to different thermal properties at the two phases is very small compared to the motions due to pressure and temperature excitations.

References

1 Carslaw, H. S.; Jaeger, J. C., 1959, Conductions of Heat in Solids, 2nd. edn, (Oxford).
2 Gjevik, B., 1972, Surface readjustment owing to subcrustal phase transition, Phy. Earth Planet. Int. **5**, 403—408.
3 Gjevik, B., 1973, The dynamical response of a subcrustal phase boundary to pressure perturbations, Phy. Earth Planet. Int. **7**, 401—408.
4 Joyner, W. B., 1967, Basalt-ecoligite transition as a cause for uplift and subsidence., J. Geophy. Res. **75**, 4977—4998.
5 Mareschal, J. C.; Gangi, A. F., 1977, A linear approximation to the solution of a one-dimensional Stefan problem and its geophysical implications, Geophy. J.R. Astr. Soc. **49**, **443**—**448**.
6 McDonald, G. J. F.; Ness, N. F., 1960, Stability of phase transitions within the Earth, J. Geophy. Res. **63**, 2173—2190.
7 O'Connel, R. J.; Wasserburg, G. J., 1967, Dynamics of a phase change boundary to changes in pressure, Rev. Geophys. **5**, 329—410.
8 O'Connel, R. J.; Wasserburg, G. J., 1972, Dynamics of submergence and uplift of a sedimentary basin underlain by phase change boundary, Rev. Geophys., **10**, 335—368.
9 Van de Lindt, W. J., 1967, Movement of the Mohorovicic discontinuity under isostatic conditions, J. Geophy. Res., **72**, 1242—1247.

Eingereicht am 2. 2. 1981

Address: Dr. T. K. De, Department of Mathematics, Burdwan University, Burdwan, West Bengal, India.

KLEINE MITTEILUNGEN

ZAMM 62, 257 – 258 (1982)

P. S. Hiremath

Diffusion of Vorticity in Maxwell Fluid

1. Introduction

The diffusion of vorticity is an important aspect of fluid dynamics. It is well known that vortices are generated at the solid boundaries in the flow field and also at the interface of two different fluids in motion. While the growth of vortices leads to turbulence, the decay of vortices ensures the sustenance of laminarity of the flow. The flow in a double array of vortices in viscous fluids was first studied by Taylor [1]. Such vortices are found to decay exponentially with time due to viscous action. Since many real fluids are non-Newtonian, the diffusion of vorticity in such fluids makes an interesting study. Gupta [2] extended the above mentioned Taylor's problem to a certain class of viscoelastic fluids represented by Walter's [3] constitutive equation. It was shown that the vortices grow with time if the spacing of the vortices is less than a critical value depending on the elastic parameter, otherwise vortices decay. The objective of the present paper is to examine the diffusion of vorticity in Maxwell fluids. In particular, the diffusion of a line vortex filament and the flow in a double array of vortices are studied. In the former case, the amount of vorticity diffused is found to be less than that in a viscous fluid initially. In the latter case, it is shown that if the ratio of the characteristic length of the fluid to the spacing of vortices is less than $1/\sqrt{8}\,\pi$, the vortices decay exponentially with time. If that ratio is greater than $1/\sqrt{8}\,\pi$, the vortices decay in a damped oscillatory manner wherein the decay rate is independent of the viscosity as well as the spacing of vortices and depends only on the relaxation time of the Maxwell fluid.

In the following analysis, it is assumed that the flow of Maxwell fluid has

(i) a characteristic Weissenberg number very much less than unity, so that invariant time derivative in the constitutive equation could be replaced by local time derivative; and

(ii) a characteristic Reynold's number very much less than unity, so that the convective terms in the governing equations of motion could be neglected.

2. Vorticity transport equation

Maxwell fluids are characterized by the following rheological equations [4]:

$$\tau_{ij} = -p\delta_{ij} + \tau'_{ij}\,, \tag{1}$$

$$\left(1 + \lambda \frac{\partial}{\partial t}\right) \tau'_{ij} = 2\mu e_{ij}\,,$$

$$e_{ij} = \frac{1}{2}\,(v_{i,j} + v_{j,i})\,,$$

where τ'_{ij} is the deviatoric stress tensor, e_{ij} the rate of strain tensor, p the pressure, λ the relaxation time, μ the coefficient of viscosity and t the time.

The two dimensional slow motion of incompressible Maxwell fluid in a cartesian co-ordinate system is governed by the equations

$$\frac{\partial u}{\partial x} + \frac{\partial v}{\partial y} = 0\,, \tag{2}$$

$$\varrho\left(1 + \lambda \frac{\partial}{\partial t}\right) \frac{\partial u}{\partial t} = -\left(1 + \lambda \frac{\partial}{\partial t}\right) \frac{\partial p}{\partial x} + \mu \nabla^2 u\,, \tag{3}$$

$$\varrho\left(1 + \lambda \frac{\partial}{\partial t}\right) \frac{\partial v}{\partial t} = -\left(1 + \lambda \frac{\partial}{\partial t}\right) \frac{\partial p}{\partial y} + \mu \nabla^2 v\,, \tag{4}$$

where

$$\nabla^2 = \frac{\partial^2}{\partial x^2} + \frac{\partial^2}{\partial y^2},$$

u and v are velocity components in x and y directions respectively and ϱ is the density of the fluid. The vorticity ζ is given by

$$\zeta = \frac{\partial v}{\partial x} - \frac{\partial u}{\partial y}, \tag{5}$$

and its transport equation can be obtained as

$$\left(1 + \lambda \frac{\partial}{\partial t}\right) \frac{\partial \zeta}{\partial t} = \nu \nabla^2 \zeta\,, \tag{6}$$

where ν is the kinematic coefficient of viscosity of the fluid. In the following sections, we obtain the solution of the equation (6) for two specific situations.

3. Line vortex filament

Now we consider the diffusion of a line vortex filament of strength K initially situated at the origin in Maxwell fluid, i.e.,

$$\zeta(x, y, 0) = \frac{K}{2}\,\delta(x)\,\delta(y)\,, \tag{7}$$

where δ is the Dirac's delta function [5]. Using the double Fourier transforms defined by

$$\bar{\zeta}(\omega_1, \omega_2, t) = \frac{1}{2\pi} \int_{-\infty}^{\infty} \int_{-\infty}^{\infty} \zeta(x, y, t)\, e^{i(\omega_1 x + \omega_2 y)}\, dx\, dy\,, \tag{8}$$

the equation (6) is transformed to

$$\lambda \frac{d^2\bar{\zeta}}{dt^2} + \frac{d\bar{\zeta}}{dt} + \nu(\omega_1^2 + \omega_2^2)\, \bar{\zeta} = 0\,. \tag{9}$$

Accordingly the initial condition (7) becomes

$$\bar{\zeta}(\omega_1, \omega_2, 0) = \frac{K}{4\pi}. \tag{10}$$

The equation (9) is analogous to the equation governing damped vibrations of a point mass in a mechanical system. This leads us to expect the diffusion of vorticity in a damped oscillatory manner in time. The second order differential equation (9) is solved subject to the initial condition (10) together with an additional condition that the solution tends to the Newtonian solution as λ tends to zero. The solution is given by

$$\bar{\zeta} = \frac{K}{4\pi}\, e^{mt}\,, \tag{11}$$

where

$$m = \frac{1}{2\lambda}\,[-1 + (1 - 4\lambda\nu(\omega_1^2 + \omega_2^2))^{1/2}] \tag{12}$$

Applying the inverse transform, we get

$$\zeta(x, y, t) = \frac{K}{8\pi^2} \int_{-\infty}^{\infty} \int_{-\infty}^{\infty} e^{mt - i(\omega_1 x + \omega_2 y)}\, d\omega_1\, d\omega_2\,. \tag{13}$$

Since the relaxation time λ is a small parameter, expanding the integrand in powers of λ and integrating, we obtain

$$\begin{aligned}\zeta(x, y, t) = {}& \frac{K}{8\pi\nu t} \exp\left(-\frac{x^2 + y^2}{4\nu t}\right) - \\ & - \frac{K\lambda}{32\pi\nu t^2} \left[D_4\left(\frac{x}{\sqrt{2\nu t}}\right) \exp\left(-\frac{x^2 + 2y^2}{8\nu t}\right) + \right. \\ & + D_4\left(\frac{y}{\sqrt{2\nu t}}\right) \exp\left(-\frac{2x^2 + y^2}{8\nu t}\right) + \\ & \left. + 2D_2\left(\frac{x}{\sqrt{2\nu t}}\right) D_2\left(\frac{y}{\sqrt{2\nu t}}\right) \exp\left(-\frac{x^2 + y^2}{8\nu t}\right)\right] + \\ & + O(\lambda^2)\,,\end{aligned} \tag{14}$$

where $D_n(Z)$ is the parabolic cylindrical function of order n [6]. The first term corresponds to the Newtonian solution [5]. We observe that the amount of vorticity diffused in Maxwell fluid is less than that in a viscous fluid by the quantity given in the second term. As time elapses, the reduction due to λ tends to zero. This can also be viewed as the effect of decaying stress relaxation as time passes. Ultimately, the situation turns out of be the same as the Newtonian one.

4. Flow in a double array of vortices in a MAXWELL fluid

Using the stream function ψ defined by

$$u = \frac{\partial\psi}{\partial y}, \qquad v = -\frac{\partial\psi}{\partial x}, \tag{15}$$

the vorticity equation takes the form

$$\left(1 + \lambda \frac{\partial}{\partial t}\right) \frac{\partial}{\partial t} \nabla^2\psi = \nu\nabla^4\psi\,. \tag{16}$$

Following TAYLOR, we assume

$$\psi = \beta \cos\frac{\pi x}{d} \cos\frac{\pi y}{d} \exp\left(-\alpha t\right) \tag{17}$$

for the flow representing the double array of vortices with assumed periodicity in both x and y directions; α and β being the constants to be determined and d being the spacing of the vortices. The constant β can be obtained by prescribing an initial condition. Substitution of (17) in (16) yields the following equation for the constant α:

$$\lambda\alpha^2 - \alpha + 2\nu\pi^2/d^2 = 0\,. \tag{18}$$

The roots of the quadratic equation (18) are

$$\alpha = \frac{1}{2\lambda}\left[1 \pm (1 - 8\Lambda^2\pi^2)^{1/2}\right], \tag{19}$$

where $\Lambda = \sqrt{\lambda\nu}/d$. Only the negative sign of the radical is considered on the requirement that (19) shall reduce to the Newtonian value, $\frac{2\nu\pi^2}{d^2}$, as λ tends to zero.

The parameter Λ represents a non-dimensional characteristic length of the MAXWELL fluid. From equation (17) and (19), we see that if $\Lambda < (1/\sqrt{8}\,\pi)$, the vortices decay exponentially with time at the rate α. If $\Lambda = (1/\sqrt{8}\,\pi)$, the vortices decay expontially with time at the rate $1/2\lambda$. If $\Lambda > (1/\sqrt{8}\,\pi)$, the vortices decay in a damped oscillatory manner with decay rate $1/2\lambda$ and frequency $(8\pi^2\Lambda^2 - 1)^{1/2}$. Thus, for $\Lambda \geqq (1/\sqrt{8}\,\pi)$, the decay rate is independent of the viscosity and the spacing of vortices and it depends only on the relaxation time of the MAXWELL fluid.

Acknowledgement

The author is indebted to Professor Dr. E. BECKER for his helpful criticism of an earlier draft of this paper.

Also, the author wishes to thank Dr. B. SIDDAPPA for many valuable discussions and Prof. P. S. MANI SUNDARAM, Principal, and Dr. V. P. MUTHUSWAMY, the Head of the Department of Mathematics, Regional Engineering College, Tiruchirapalli, for the encouragement.

References

1 TAYLOR, G. I., On the Decay of Vortices in a Viscous Fluid, Phil. Mag. 46, 671–674 (1923).
2 GUPTA, A. S., Growth or decay of vortices in a Viscoelastic Liquid, Jour. Math. Phys. Sci. 6, 263–270 (1972).
3 WALTERS, K., Second Order Effects in Elasticity, Plasticity and Fluid Dynamics, Pergamon Press, (1964).
4 EIRICH, F. R., Rheology, Vol. I, Academic Press, (1956).
5 SNEDDON, I. N., FOURIER Transforms, McGraw-Hill (1951).
6 ERDELYI, A., Tables of Integral Transforms, Vol. I, McGraw-Hill, (1954).

Eingereicht am 17. 6. 1981

Anschrift: PRAKASH S. HIREMATH, Department of Mathematics, Regional Engineering College, Tiruchirapalli-620015. (India)

ZAMM 62, 258–261 (1982)

L. F. HENDERSON

Exact Expressions for Shock Reflexion Transition Criteria in a Perfect Gas

This paper presents exact formulae for the three transition criteria between regular and Mach reflexion for a plane shock wave propagating in a perfect gas. The formulae for the normal shock point, and the shock detachment point have been given by VON NEUMANN (1943), but not in a form convenient for modern usage. They have been reworked for presentation here. The sonic point formula has not been given before in any form and is also presented here. VON NEUMANN proposed that the boundary which separates weak from strong shocks, should be at the particular condition where the normal shock and detachment points coincide. A formula is given here for this condition. However recent experiments by HENDERSON and SIEGENTHALER (1980) indicate that it may be more appropriate to take the coincidence of the normal shock and sonic points as the separation condition. This matter is discussed and some numerical data are presented. Other formulae relevant to the problem are also presented here for the first time, including the condition for a normal reflected shock in Mach reflexion.

1. Introduction

VON NEUMANN (1943) completely solved the problem of a plane shock wave experiencing regular reflexion RR off a rigid plane surface in a perfect gas. He succeeded in reducing the equations to a single quadratic. This equation can be reworked so that the dependent variable is the shock Mach number M_r of the reflected shock r, and its coefficients are functions only of $(\gamma, \xi_i, \omega_0)$. Here $\gamma \equiv C_p/C_v$, is the ratio of specific heats, $\xi_i \equiv P_0/P_1$, is the inverse shock strength, and ω_0 is the angle of incidence of the incident shock i; the subscripts 0, 1, refer to conditions upstream and downstream of i respectively, figure 1. If (γ, ξ_i) are

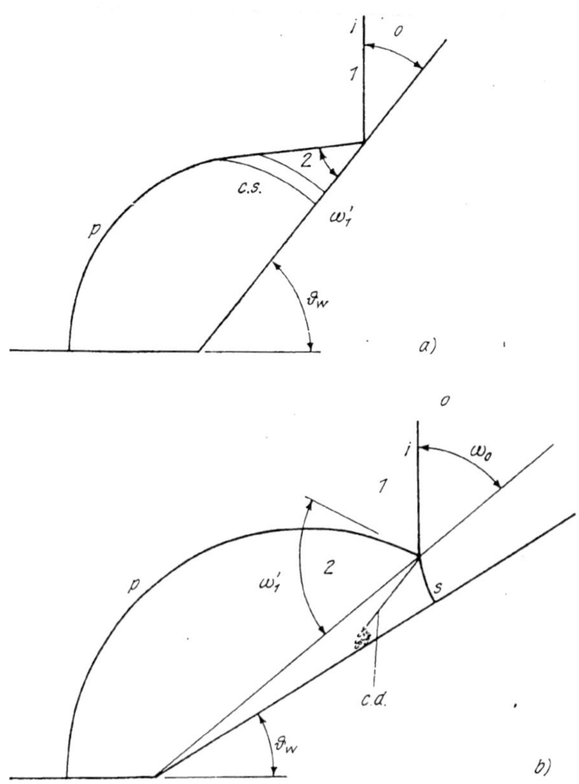

Fig. 1. Nomenclature (a) Regular reflexion RR; (b) Irregular (Mach) reflexion MR. i, incident shock; r, reflected shock; s, Mach stem; cd, contact discontinuity; $c.s.$, corner signal

held constant while ω_0 is allowed to increase then eventually RR will give way to Mach reflexion MR. The solution for the confluence of the three shocks in MR has been given by HENDERSON (1964). He obtained a single polynomial equation of tenth degree, and then found two (physically trivial) double roots so that on division the equation was reduced finally to one of sixth degree.

Three criteria have been proposed for determining the transition condition RR $\rightleftharpoons$ MR, these are often referred to as the normal shock point $\omega_0 = \omega_N$, the sonic point $\omega_0 = \omega_0^*$, and the shock detachment point $\omega_0 = \omega_e$, the latter is also called the extreme point. Their ordering is as follows $\omega_N < \omega_0^* < \omega_e$. The question as to which criterion determines transition has been the subject of controversy in recent years, HENDERSON and LOZZI (1975, 1979), BAZHENOVA et al. (1976), HORNUNG and KYCHAKOFF (1977), HEILIG (1978), BEN-DOR and GLASS (1979, 1980), HORNUNG et al. (1979), HENDERSON and SIEGENTHALER (1980), but we do not enter the controversy here. The purpose of this paper is to present closed form expressions, in a format convenient for modern usage, of the VON NEUMANN regular reflexion solution, of all three transition criteria, together with some associated formulae and particularly those concerned with the weak-strong shock boundary. The VON NEUMANN formulae have not been published before in the form given here, and the sonic criterion has not been previously published in any form nor has the condition for the reflected shock to be normal to the flow in Mach reflexion.

2. The Formulae

2.1. Regular Reflexion

The theory of the regular reflexion of a plane shock wave i, in a perfect gas can be reduced to the following quadratic,

$$M_r^4 - C_1 M_r^2 + C_0 = 0\,, \tag{2.01}$$

so that,

$$M_r^2 = \tfrac{1}{2}\,[C_1 \pm (C_1^2 - 4C_0)^{1/2}]\,, \tag{2.02}$$

where,

$$C_0 \equiv 1 - \frac{4(1-\xi_i)^2 \cos^2\omega_0}{[(\gamma-1)+(\gamma+1)\,\xi_i]^2} + \frac{[(\gamma+1)+(\gamma-1)\,\xi_i]^2 \cot^2\omega_0}{[(\gamma-1)+(\gamma+1)\,\xi_i]^2}\,, \tag{2.03a}$$

$$C_1 \equiv 2 + \frac{2(1-\xi_i)^2\cos^2\omega_0}{[(\gamma-1)+(\gamma+1)\,\xi_i]} + \frac{[(\gamma+1)+(\gamma-1)\,\xi_i]^2\cot^2\omega_0}{2\gamma[(\gamma-1)+(\gamma+1)\,\xi_i]}\,. \tag{2.03b}$$

Here the negative branch refers to the weak solution, that is, the one for which the flow downstream of the reflected shock r is supersonic $M_2 > 1$ over most of the range of ω_0, and as far as is known this is always the solution which is physically relevant. The positive branch refers to the strong solution, that is the one where the flow downstream of r is always subsonic $M_2 < 1$, this solution probably corresponds to a flow which is unstable HENDERSON and ATKINSON (1976).

2.2. The Transition Criteria

2.2.1. The normal shock point criterion, $\omega_0 = \omega_N$. At this condition the Mach stem is everywhere normal to the flow, but because it is at transition the stem must also be of zero length. The equation for the criterion may be derived by replacing the boundary condition for RR by the condition for a normal shock. Then one obtains,

$$C_2 \sin^4\omega_N - C_1 \sin^2\omega_N + C_0 = 0\,, \tag{2.04}$$

so that

$$\sin^2\omega_N = (2C_2)^{-1}\,[C_1 \pm (C_1^2 - 4C_0C_2)^{1/2}]\,, \tag{2.05}$$

where,

$$C_0 \equiv [(\gamma+1)+(\gamma-1)\,\xi_i]^2\,[(\gamma-1)+(\gamma+1)\,\xi_i]\,, \tag{2.06a}$$

$$C_1 \equiv [(\gamma+1)+(\gamma-1)\,\xi_i]\,[(\gamma^3+\gamma^2+3\gamma-1)+(\gamma-1)\times(\gamma^2+3\gamma-2)\,\xi_i+(\gamma+1)(3\gamma-1)\,\xi_i^2]\,, \tag{2.06b}$$

$$C_2 \equiv 2\gamma(1-\xi_i)\,[(\gamma+1)^2+2(\gamma-1)\,\xi_i-(\gamma+1)(\gamma-1)\,\xi_i^2]\,. \tag{2.06c}$$

The solution is given by the negative branch of (2.05). In the limit $\xi_i \to 0$, we have,

$$C_0 \equiv (\gamma+1)(\gamma-1); \qquad C_1 \equiv (\gamma^3+\gamma^2+3\gamma-1);$$
$$C_2 \equiv 2\gamma(\gamma+1)\,, \tag{2.07}$$

and this gives for example $\omega_N = 21 \cdot 769°$ when $\gamma = 7/5$. An important property of this criterion is that the pressure downstream of the shock system changes continuously through transition provided that this is induced by continuous changes in (γ, ξ_i) or the geometry. Hence the flow always remains in mechanical equilibrium. But the pressure changes discontinuously for the other criteria so there is no such equilibrium for them.

2.2.2. The sonic point criterion, $\omega_0 = \omega_0^*$. At this condition the flow downstream of the reflected shock r is sonic, $M_2 = 1$. The criterion can be derived by eliminating M_r^2 between (2.01) and the condition for sonic flow downstream of an oblique shock, and this is given by AMES (1953). For present purposes the AMES equation has been reworked into the following form,

$$M_r^4 - \frac{(\gamma+1)\,M_1^2 - (3-\gamma)}{2\gamma}\,M_r^2 - \frac{1}{\gamma} = 0\,, \tag{2.08}$$

where M_1 is the Mach number of the flow between i and r, and

$$M_1^2 = \frac{[(\gamma+1)+(\gamma-1)\,\xi_i]^2 - 4\gamma(1-\xi_i)^2\sin^2\omega_0^*}{2\gamma[(\gamma-1)+(\gamma+1)\,\xi_i]\sin^2\omega_0^*}\,, \tag{2.09}$$

(2.08) and (2.09) are equivalent to the AMES result. Now eliminating M_1^2 between them and then M_r^2 between the new equation and equation (1), we get after a notably tedious calculation,

$$C_5\sin^{10}\omega_0^* + C_4\sin^8\omega_0^* + C_3\sin^6\omega_0^* + C_2\sin^4\omega_0^* + C_1\sin^2\omega_0^* + C_0 = 0\,, \tag{2.10}$$

where

$$C_0 \equiv (\gamma-1)\,[(\gamma+1)+(\gamma-1)\,\xi_i]^6\,, \tag{2.11a}$$

$$C_1 \equiv -2[(\gamma+1)+(\gamma-1)\,\xi_i]^4\,\{(\gamma-1)(5\gamma^2+2\gamma+1) - 2(\gamma^3-2\gamma^2+4\gamma-1)\,\xi_i + (\gamma-1)(\gamma^2-8\gamma+1)\,\xi_i^2\}\,, \tag{2.11b}$$

$$\begin{aligned}C_2 \equiv\;& 2[(\gamma+1)+(\gamma-1)\,\xi_i]^2\,\{(\gamma-1)(4\gamma^4+23\gamma^3+23\gamma^2-3\gamma+1) + 2(3\gamma^5-22\gamma^4-22\gamma^3+20\gamma^2-13\gamma+2)\,\xi_i +\\ &+ 6(\gamma-1)(-4\gamma^3+3\gamma^2-8\gamma+1)\,\xi_i^2 + 2(-\gamma^5+12\gamma^4-32\gamma^3+58\gamma^2-23\gamma+2)\,\xi_i^3 +\\ &+ (\gamma-1)(-15\gamma^3+51\gamma^2-13\gamma+1)\,\xi_i^4\}\,,\end{aligned} \tag{2.11c}$$

$$\begin{aligned}C_3 \equiv\;& 4\gamma\{-2(\gamma-1)(\gamma+1)^2(3\gamma^2+6\gamma-1) + (\gamma+1)(3\gamma^4+80\gamma^3+70\gamma^2-72\gamma+15)\,\xi_i +\\ &+ (33\gamma^5+31\gamma^4-214\gamma^3-98\gamma^2+165\gamma-45)\,\xi_i^2 + 2(11\gamma^5-65\gamma^4+22\gamma^3+110\gamma^2-145\gamma+35)\,\xi_i^3 +\\ &+ 4(\gamma-1)(-3\gamma^4+4\gamma^3+28\gamma^2-60\gamma+15)\,\xi_i^4 + (-9\gamma^5+47\gamma^4-194\gamma^3+326\gamma^2-165\gamma+27)\,\xi_i^5 +\\ &+ (\gamma-1)(\gamma^4-40\gamma^3+82\gamma^2-32\gamma+5)\,\xi_i^6\}\,,\end{aligned} \tag{2.11d}$$

$$\begin{aligned}C_4 \equiv\;& 16\gamma^2(1-\xi_i)^2\,\{(\gamma-1)(\gamma+1)^2 - 2(\gamma+1)(\gamma+3)\times(3\gamma-1)\,\xi_i - (\gamma-1)(13\gamma^2+12\gamma-13)\,\xi_i^2 -\\ &- 4(\gamma^3-\gamma^2+7\gamma-3)\,\xi_i^3 + 2(\gamma^3-10\gamma^2+7\gamma-2)\,\xi_i^4\}\,,\end{aligned} \tag{2.11e}$$

$$C_5 \equiv 64\gamma^3(1-\xi_i)^4\,\xi_i[(\gamma+1)+(\gamma-1)\,\xi_i]\,. \tag{2.11f}$$

Equation (2.10) is badly conditioned at $\xi_i = 0$ and 1, and also at $\xi_i = -(\gamma+1)/(\gamma-1)$ but this last value is of no physical significance. In the limit $\xi_i \to 0$, (2.11) become,

$$C_0 \equiv (\gamma+1)^4; \qquad C_1 \equiv -2(\gamma+1)^2(5\gamma^2+2\gamma+1);$$
$$C_2 \equiv 2(4\gamma^4+23\gamma^3+23\gamma^2-3\gamma+1);$$
$$C_3 \equiv -8\gamma(3\gamma^2+6\gamma-1); \qquad C_4 \equiv 16\gamma^2; \qquad C_5 = 0\,, \tag{2.12}$$

and with (2.10) and $\gamma = 7/5$ this gives for example $\omega_0^* = 39 \cdot 910°$. In the other limit $\xi_i \to 1$, (2.10) and (2.11) together reduce to,

$$2\sin^6\omega_0^* + (\gamma - 5)\sin^4\omega_0^* + 2(2-\gamma)\sin^2\omega_0^* + (\gamma - 1) \equiv$$

$$\equiv (\sin^2\omega_0^* - 1)^2\left(\sin^2\omega_0^* + \frac{\gamma - 1}{2}\right) = 0\,. \quad (2.13)$$

This is the acoustic limit and the only result of physical significance is that $\omega_0^* = \pi/2$ at $\xi_i = 1$, which is as it should be.

2.2.3. The shock detachment criterion, $\omega_0 = \omega_e$. This is also called the extreme angle criterion, and it may be derived from the condition that the two roots in equation (2.02) become coincident at $\omega_0 = \omega_e$. Then setting the discriminant equal to zero we get,

$$C_3\sin^6\omega_e + C_2\sin^4\omega_e + C_1\sin^2\omega_e + C_0 = 0\,, \quad (2.14)$$

where,

$$C_0 \equiv -[(\gamma+1) + (\gamma-1)\,\xi_i]^4\,, \quad (2.15a)$$

$$C_1 \equiv [(\gamma+1) + (\gamma-1)\,\xi_i]^2\,\{(9\gamma^2 + 2\gamma + 1) - 2(3\gamma^2 - 4\gamma + 1)\,\xi_i + (\gamma^2 - 10\gamma + 1)\,\xi_i^2\}\,, \quad (2.15b)$$

$$C_2 \equiv -8\gamma(1-\xi_i)^2\,\{(3\gamma^2 + 4\gamma - 1) + 2(\gamma^2+1)\,\xi_i - (\gamma^2 - 4\gamma + 1)\,\xi_i^2\}\,, \quad (2.15c)$$

$$C_3 \equiv 16\gamma^2(1-\xi_i)^4\,. \quad (2.15d)$$

When $\xi_i \to 0$, the coefficients reduce to,

$$C_0 \equiv -(\gamma+1)^4; \qquad C_1 \equiv (\gamma+1)^2\,(9\gamma^2 + 2\gamma + 1);$$

$$C_2 \equiv -8\gamma(3\gamma^2 + 4\gamma - 1); \qquad C_3 \equiv 16\gamma^2\,, \quad (2.16)$$

and with $\gamma = 7/5$ we get $\omega_e = 39 \cdot 971°$ when $\xi_i = 0$. At the acoustic limit $\xi_i \to 1$, (2.14) and (2.15) reduce to $\sin^2\omega_e - 1 = 0$, so that $\omega_e = \pi/2$. When $\omega_0 > \omega_e$ the roots are complex and according to the theory regular reflexion is then impossible.

2.3. Conditions for the Coincidence of the Criteria

For certain values of (γ, ξ_i), the normal shock point coincides with the detachment point $\omega_N = \omega_e$, and VON NAUMANN proposed that this should be the boundary between weak and strong shocks. The condition can be found by eliminating $\sin^2\omega_0$ between (2.04) and (2.14). This gives,

$$C_4\xi_i^4 + C_3\xi_i^3 + C_2\xi_i^2 + C_1\xi_i + C_0 = 0\,, \quad (2.17)$$

where

$$C_0 \equiv -\gamma^4 + 6\gamma^2 + 24\gamma + 3\,,$$

$$C_1 \equiv -4(\gamma+1)\,(\gamma^2 + 8\gamma + 3)\,,$$

$$C_2 \equiv 2(\gamma+3)\,(\gamma+1)\,(\gamma^2+3)\,, \quad (2.18)$$

$$C_3 \equiv 12(\gamma-1)\,(\gamma+1)^2\,,$$

$$C_4 \equiv (3-\gamma)\,(\gamma+1)^3\,.$$

When $\gamma = 7/5$ we get $\xi_i = 0.43317$, and then $\omega_0 = \omega_e = \omega_N = 41.412°$. In their recent experiments with shocks diffracting over concave corners, HENDERSON and SIEGENTHALER (1980) found that the expansion wave commonly called the corner signal, was always able to overtake ("catch-up") every part of the reflected shock r and the Mach stem s in the case of weak Mach reflexion, but this was not always true for strong Mach reflexion. Now because the corner signal travels everywhere at the local speed of sound, the experiments indicate that a more appropriate boundary between weak and strong shocks is the coincidence of the sonic point with the normal shock point $\omega_0^* = \omega_N$. In order to obtain the formula we have to eliminate $\sin^2\omega_0$ between equations (2.04) and (2.10), and although straightforward it is a daunting task, given the complexity of the coefficients in (2.10). We have not done this, but it is simple enough to obtain some numerical data. Some data for both conditions are compared in Table 1.

2.4. Conditions for a Normal Reflected Shock in Mach Reflexion

For given (γ, ξ_i) there exists a certain wave angle $\omega_0 = \omega_P$ say, for which the reflected shock wave becomes locally normal to the flow in Mach reflexion. This condition is often of theoretical interest. We have that $M_r = M_1$ and $\omega_1 = \pi/2$, where ω_1 is the wave angle of r with respect to the flow between i and r. Let

$$a \equiv (\gamma-1)\,\xi_i + (\gamma+1)\,,$$

$$b \equiv (\gamma+1)\,\xi_i + (\gamma-1)\,,$$

$$c \equiv 4\gamma\xi_i^2 - (\gamma+1)\,(\gamma-1)\,\xi - (\gamma+1)^2\,,$$

then the condition is,

$$c_4\sin^8\omega_P + c_3\sin^6\omega_P + c_2\sin^4\omega_P + c_1\sin^2\omega_P + c_0 = 0 \quad (2.19)$$

where,

$$c_4 \equiv 4\gamma(1-\xi_i)\,\{(\gamma+1)^2\,b^3\xi_i^2[(3-\gamma)\,\xi_i - (\gamma+1)] - (\gamma+1)\,(3-\gamma)\,b^2c\xi_i^2(1+\xi_i) + bc^2[4\xi_i^2 + (\gamma+1)\,\xi_i + (\gamma+1)] - c^3(1+\xi_i)\}\,, \quad (2.19a)$$

$$c_3 \equiv a\{(\gamma+1)^2\,b^3\xi_i[2(\gamma^2 - 6\gamma + 1)\,\xi_i^2 + (\gamma^2 + 10\gamma - 3)\,\xi_i - (\gamma-1)\,(\gamma+1)] - (\gamma+1)\,b^2[4\gamma(3-\gamma)\,a(1-\xi_i)\,(1+\xi_i)\,\xi_i^2 + c[2(\gamma^2 - 6\gamma + 1)\,\xi_i^3 - (\gamma^2 - 4\gamma + 3)\,\xi_i + (\gamma+1)^2]] + bc[8\gamma a(1-\xi_i)\,[4\xi_i^2 + (\gamma+1)\,\xi_i + (\gamma+1)] - c\xi_i[(\gamma^2 + 10\gamma - 3)\,\xi_i - (\gamma-1)\,(3-\gamma)]] + ac^2[c - 12\gamma(1-\xi_i)\,(1+\xi_i)]\}\,, \quad (2.19b)$$

$$c_2 \equiv a^2\{(\gamma+1)^2\,b^3\xi_i[-(3\gamma-1)\,\xi_i + (\gamma-1)] + (\gamma+1)\,b^2[-a[2(\gamma^2 - 6\gamma + 1)\,\xi_i^3 - (\gamma^2 - 4\gamma + 3)\,\xi_i + (\gamma+1)^2] + c[(3\gamma-1)\,\xi_i^2 + (\gamma+1)]] + ab[4\gamma a(1-\xi_i)\,[4\xi_i^2 + (\gamma+1)\,\xi_i + (\gamma+1)] - 2c\xi_i[(\gamma^2 + 10\gamma - 3)\,\xi_i - (\gamma-1)\,(3-\gamma)] - c^2] + 3a^2c[c - 4\gamma(1-\xi_i)\,(1+\xi_i)]\}\,, \quad (2.19c)$$

$$c_1 \equiv a^4\{(\gamma+1)\,b^2[(3\gamma-1)\,\xi_i^2 + (\gamma+1)] - ab[2c + \xi_i[(\gamma^2 + 10\gamma - 3)\,\xi_i - (\gamma-1)\,(3-\gamma)]] + a^2[3c - 4\gamma(1-\xi_i)\,(1+\xi_i)]\}\,, \quad (2.19d)$$

$$c_0 \equiv a^7(a-b)\,. \quad (2.19e)$$

2.5. Conditions for Sonic Flow between the Incident and Reflected Shock in Mach Reflexion

This condition has been given by HENDERSON and SIEGENTHALER, it is reworked here in terms of (γ, ξ_i) and the particular wave angle at which it occurs $\omega_0 = \omega_M$ say. We now have that $M_1 = 1 = M_r$ and,

$$\sin^2\omega_M = \frac{[(\gamma+1) + (\gamma-1)\,\xi_i]^2}{2\gamma[-2\xi_i^2 + (\gamma+1)\,\xi_i + (\gamma+1)]}\,. \quad (2.20)$$

At this condition it is also true that $\omega_1 = \pi/2$, so it follows at once that equation (2.20) is a root of equation (2.19). Hence (2.19) can be reduced to a cubic with the help of (2.20) but it would be a tedious task to divide the root out. However, this can be done numerically quite easily on a computer.

Table 1

γ	7/5	1.402	5/3
ξ_i	0.43317	0.43262	0.36385
$\omega_N = \omega_e$	41.412	41.392	38.833
M_i	1.4566	1.4572	1.5488
M_r	1.4753	1.4768	1.5301
ω_1'	50.4126	50.5001	49.8081
M_2	0.9213	0.9196	0.9246
γ	7/5	1.402	5/3
ξ_i	0.375311	0.374744	0.303750
$\omega_N = \omega_e^*$	40.374	40.353	37.680
M_i	1.5578	1.5586	1.6834
M_r	1.5032	1.5037	1.5630
ω_1'	44.4228	44.4206	44.1594
M_2	1.0000	1.0000	1.0000

Comparison of the two criteria $\omega_N = \omega_e$ and $\omega_N = \omega_0^*$ as the criterion to distinguish between weak and strong shock waves.

Acknowledgement: This work is supported by the Australian Research Grants Committee.

References

AMES Research Staff 1953, Equations, Tables and Charts for Compressible Flow, Nat. Adv. Comm. for Aero. Rept. 1135 Wash. D.C.
BAZHENOVA, T. V., FOKEEV, V. P. & GVOZDEAVA, L. G. 1976, Acta Astron. **3**, 131–140.
BEN-DOR, G. & GLASS, I. I. 1979, J. Fluid Mech. **92**, 459–496.
BEN-DOR, G. & GLASS, I. I. 1980, J. Fluid Mech. **96**, 735–756.
HEILIG, W. H. 1978, Shock Tube and Shock Wave Research, Proc. Eleventh Intern. Symp. on Shock Tubes & Waves, Univ. Wash. Press, Seattle, 288–295.
HENDERSON, L. F. 1964, Aero. Quart. **15**, 181–197.
HENDERSON, L. F. & ATKINSON, J. D. 1976, J. Fluid Mech., **75**, 751–764.
HENDERSON, L. F. & LOZZI, A. 1975, J. Fluid Mech., **68**, 139–155.
HENDERSON, L. F. & LOZZI, A. 1979, J. Fluid Mech., **94**, 541–560.
HENDERSON, L. F. & SIEGENTHALER, A. 1980, Proc. Roy. Soc. Lond. **A369**, 537–555.
HORNUNG, H. G. & KYCHAKOFF, G. 1978, Shock Tube and Shock Wave Research, Proc. Eleventh Intern. Symp. on Shock Tubes, Univ. of Wash. Press, 296 to 302.
HORNUNG, H. G., OERTEL, H. & SANDEMAN, R. J. 1979, J. Fluid Mech., **90**, 541–560.
NEUMANN, J. VON 1943, Oblique Reflexion of Shock Waves, JOHN VON NEUMANN, Collected Works VI, p. 238, Pergamon Press.

Eingereicht am 5. 3. 1981

Address: Dr. L. F. HENDERSON, School of Mechanical and Aeronautical Engineering, University of Sydney, N.S.W. 2006, Australia.

ZAMM 62, 261–264 (1982)

B. M. SRIVASTAVA / R. P. AGARWAL / V. P. SRIVASTAVA

Unsteady Flow of Suspension through a Rectangular Pipe

1. Introduction

The flow behavior of many substances such as suspensions, polymers, polycrystalline aggregates, granular media, bubbly liquids, liquid crystals, fluid filled porous elastic solids, composite elastic solids, alloys, etc. cannot be explained fully if one regards them as single continua. In order to describe adequately the nature of such substances, it seems to be important and necessary to assume that the material consists of more than one constituent. The theoretical study of this system of fluids has been very useful in understanding such phenomena as sedimentation, fluidisation, combustion, environmental pollution, etc. Lately, interest is developed in applying this theory to hydrodynamics of biological systems. Because it provides an improved understanding of such diverse subjects, for example diffusion of proteins, swimming of microorganisms, particle deposition in respiratory tract, and to describe flow of blood in the vessels of small diameter. A more fundamental and rigorous theoretical approach of modern phase of the continuum theory of mixture can be found in the treatises of TRUESDELL [1] and TRUESDELL and TOUPIN [2] and in papers by ERINGEN and INGRAM [3], GREEN and NAGHDI [4], and MILLS [5, 6]. ATKIN and GRAINE [7] recently derived the fundamental equations of a continuum theory of mixture and gave the historical development of the subject. In a subsequent paper [8], they applied the basic theory for different types of binary chemically reacting and non-reacting mixture of fluids. WILHELM and VAN DER WERFF [9] studied binary mixture of Newtonian fluids in a pipe. In addition, number of authors have worked on multiphase or two-phase flows from different view points, such of the work being concerned with suspension of solid particles in fluids. An extensive review of the theory of solid-fluid mixture has been given by MARBLE [10, 11] and SOO [12]. These studies include the work of DAVIDSON and HARRISON [13], MURRY [14, 15, 16], BRENNER [17], DIGIOVANNI and LEE [18], DREW [19–23] and PEDDISON [24].

The research reported here is concerned with the study of unsteady flow of a binary mixture of chemically inert, incompressible Newtonian fluids through a rigid tube of rectangular cross-section under the influence of arbitrary time varying pressure gradient. Each of the constituents of the mixture is described by a modified Navier-Stokes equation, in which dynamic coupling of different fluids is due to intercomponent friction forces which satisfy the action-reaction principle. Velocity of both the fluids are explicitly determined for various particular pressure gradients of physical importance. Finally, to gain insight in patterns of flow, numerical study of the results has been made in detail using digital computer.

2. Formulation

We consider the flow of a mixture of fluid-particle system through a rigid tube of rectangular cross-section of dimensions $x = \pm a$ and $y = \pm b$. The particles have been considered as neutrally bount, rigid small spherical shaped. Particle-particle interaction and Brownian motion have been neglected. The flow is axisymmetric and is generated by the application of the pressure gradient which is an arbitrary function of time. If $u(x, y, t)$ and $v(x, y, t)$ denote the velocity of fluid and particulate phase respectively, then the use of the following dimensionless variables

$$x' = \frac{x}{a}, \quad y' = \frac{y}{a}, \quad z' = \frac{z}{a}, \quad p' = \frac{Pa^2}{\varrho v^2},$$

$$t' = \frac{tv}{a^2}, \quad u' = \frac{ua}{v}, \quad v' = \frac{va}{v}$$

reduces the appropriate equations (neglecting body forces and body couples) describing the flow of fluid particle system after dropping dashes as (Soo [12])

$$\frac{\partial u}{\partial t} = -\varphi \frac{\partial p}{\partial x} + \left(\frac{\partial^2 u}{\partial x^2} + \frac{\partial^2 u}{\partial y^2}\right) + \beta(v - u), \tag{1}$$

$$\frac{\partial v}{\partial t} = -\varphi' \frac{\partial p}{\partial x} + \left(\frac{\partial^2 v}{\partial x^2} + \frac{\partial^2 v}{\partial y^2}\right) + \gamma(u - v), \tag{2}$$

where $\varphi = 1 - K_m \frac{\varrho_p}{\bar{\varrho}_p}$, $\varphi' = \frac{\varrho}{\bar{\varrho}_p}$, $\beta = K_m \frac{\varrho_p \gamma}{\varrho}$, $\gamma = \frac{Fa^2}{v}$; μ, μ_p and ϱ, ϱ_p are the viscosity and density of two phases respectively. $\bar{\varrho}_p$ is the density of the material of the particles, K_m is the effectiveness constant. Further, the kinematic viscosities of the two phases are assumed to be the same i.e.

$$\frac{\mu}{\varrho} = \frac{\mu_p}{\varrho_p} = \nu \text{ (constant)}.$$

In view of the symmetric condition only the flow in the region $x \geqq 0$, $y \geqq 0$ is considered. Accordingly, the initial and boundary conditions are

$$u = v = 0 \quad \text{at} \quad t = 0 \tag{3}$$

$$\left.\begin{aligned} u(1, y; t) &= 0 \\ v(1, y; t) &= 0 \end{aligned}\right\} 0 \leqq y \leqq h; \qquad \frac{\partial u}{\partial x} = \frac{\partial v}{\partial x} = 0 \text{ at } x = 0, \tag{4}$$

$$\left.\begin{aligned} u(x, h; t) &= 0 \\ v(x, h; t) &= 0 \end{aligned}\right\} 0 \leqq x \leqq 1; \qquad \frac{\partial u}{\partial y} = \frac{\partial v}{\partial y} = 0 \text{ at } y = 0 \tag{5}$$

with $h = \frac{b}{a}$.

3. Analysis

The solution of the equation (1) and (2) are obtained by the consecutive use of Laplace and cosine transforms, defined as

$$\tilde{f}(n, m; t) = \int_0^\infty e^{-st} \, dt \int_0^1 \int_0^h f(x, y; t) \cos q_n x \cos q_m y \, dx \, dy, \tag{6}$$

where $q_n = \frac{(2n + 1)\pi}{2}$, $q_m = \frac{(2m + 1)\pi}{2h}$; tilde and bar stand for cosine and Laplace transform respectively.

The inverse of the joint transform (6) is given by

$$f(x, y; t) = \frac{1}{2\pi i} \int_{\alpha - i\infty}^{\alpha + i\infty} e^{st} \left[\frac{4}{h} \sum_{n=0}^{\infty} \sum_{m=0}^{\infty} \bar{\tilde{f}}(n, m; t) \cos q_n x \cos q_m y\right] ds, \tag{7}$$

with $\alpha \geqq 0$.

Application of the joint transform defined in (6) along with the foregoing initial and boundary conditions (3–5) reduce the equations (1) and (2) to

$$s\bar{\tilde{u}} = \frac{(-1)^{n+m}}{q_n q_m} \varphi \bar{f}(s) - (q_n^2 + q_m^2) \bar{\tilde{u}} + \beta(\bar{\tilde{v}} - \bar{\tilde{u}}), \tag{8}$$

$$s\tilde{\bar{v}} = \frac{(-1)^{n+m}}{q_n q_m} \varphi' \bar{f}(s) - (q_n^2 + q_m^2)\, \tilde{\bar{v}} + \gamma(\tilde{\bar{u}} - \tilde{\bar{v}}) \tag{9}$$

where $-\dfrac{\partial p}{\partial z} = f(t)$ and $\bar{f}(s)$ stands for the Laplace transform of $f(t)$.

Solving equations (8) and (9) for u and v, using the inverse transform defined in (7) and the convolution theorem, the expressions for the velocity of two phases are given as

$$u(x, y; t) = \frac{4}{h} \sum_{n=0}^{\infty} \sum_{m=0}^{\infty} \frac{(-1)^{n+m}}{q_n q_m(\beta + \gamma)} \left[\int_0^t f(t - \eta) \times \right.$$
$$\left. \times \{(\beta\varphi' \gamma\varphi)\, e^{\alpha_1 \eta} + \beta(\varphi - \varphi')\, e^{\alpha_2 \eta}\}\, d\eta \right] \cos q_n x \cos q_m y \tag{10}$$

and

$$v(x, y; t) = \frac{4}{h} \sum_{n=0}^{\infty} \sum_{m=0}^{\infty} \frac{(-1)^{n+m}}{q_n q_m(\beta + \gamma)} \left[\int_0^t f(t - \eta) \times \right.$$
$$\left. \times \{(\beta\varphi' + \gamma\varphi)\, e^{\alpha_1 \eta}\, \gamma(\varphi' - \varphi)\, e^{\alpha_2 \eta}\}\, d\eta \right] \cos q_n x \cos q_m y, \tag{11}$$

where

$$\binom{\alpha_1}{\alpha_2} = -0.5\, [\beta + \gamma + 2(q_n^2 + q_m^2) \mp \{[\beta + \gamma + 2(q_n^2 + q_m^2)]^2 -$$
$$- 4(q_n^2 + q_m^2)\, (\beta + \gamma + q_n^2 + q_m^2)\}^{1/2}]\,. \tag{12}$$

4. Particular Cases

(i) Pressure gradient is constant.

Substituting $f(t) = C$ in equations (10) and (11), one gets the expressions for the velocity of two phases as

$$u = \frac{4C}{h} \sum_{n=0}^{\infty} \sum_{m=0}^{\infty} \frac{(-1)^{n+m}}{q_n q_m} \left[\frac{\beta\varphi' + (\gamma + q_n^2 + q_m^2)\, \varphi}{\alpha_1 \alpha_2} + \right.$$
$$\left. + \frac{1}{\beta + \gamma} \left\{ \frac{(\beta\varphi' + \gamma\varphi)\, e^{\alpha_1 t}}{\alpha_1} + \frac{\beta(\varphi - \varphi')\, e^{\alpha_2 t}}{\alpha_2} \right\} \right] \cos q_n x \cos q_m y \tag{13}$$

and

$$v = \frac{4C}{h} \sum_{n=0}^{\infty} \sum_{m=0}^{\infty} \frac{(-1)^{n+m}}{q_n q_m} \left[\frac{\gamma\varphi + (\beta + q_n^2 + q_m^2)\, \varphi'}{\alpha_1 \alpha_2} + \right.$$
$$\left. + \frac{1}{\beta + \gamma} \left\{ \frac{(\gamma\varphi + \beta\varphi')\, e^{\alpha_1 t}}{\alpha_1} + \frac{\gamma(\varphi' - \varphi)\, e^{\alpha_2 t}}{\alpha_2} \right\} \right] \cos q_n x \cos q_m y\,. \tag{14}$$

(ii) When pressure gradient is an oscillatory function of time i.e. $f(t) = C \sin \omega t$, then the velocities of two phases become

$$u = \frac{4C}{h} \sum_{n=0}^{\infty} \sum_{m=0}^{\infty} \frac{(-1)^{n+m}}{q_n q_m} \left[\frac{\xi \sin \omega t + \eta \cos \omega t}{(\alpha_1^2 + \omega^2)\, (\alpha_2^2 + \omega^2)} + \right.$$
$$\left. + \frac{\omega}{\beta + \gamma} \left\{ \frac{(\beta\varphi' + \gamma\varphi)\, e^{\alpha_1 t}}{\alpha_1^2 + \omega^2} + \frac{\beta(\varphi - \varphi')\, e^{\alpha_2 t}}{\alpha_2^2 + \omega^2} \right\} \right] \cos q_n x \cos q_m y \tag{15}$$

and

$$v = \frac{4C}{h} \sum_{n=0}^{\infty} \sum_{m=0}^{\infty} \frac{(-1)^{n+m}}{q_n q_m} \left[\frac{\xi' \sin \omega t + \eta' \cos \omega t}{(\alpha_1^2 + \omega^2)\, (\alpha_2^2 + \omega^2)} + \right.$$
$$\left. + \frac{\omega}{\beta + \gamma} \left\{ \frac{(\gamma\varphi + \beta\varphi')\, e^{\alpha_1 t}}{\alpha_1^2 + \omega^2} + \frac{\gamma(\varphi' - \varphi)\, e^{\alpha_2 t}}{\alpha_2^2 + \omega^2} \right\} \right] \cos q_n x \cos q_m y \tag{16}$$

with

$$\xi = \{\beta(\varphi' - \varphi) - \alpha_2 \varphi\}\, \{(\alpha_1 \alpha_2 - \omega^2) - \omega^2 \varphi(\alpha_1 + \alpha_2)\}\,,$$
$$\xi' = \{\gamma(\varphi - \varphi') - \alpha_2 \varphi'\}\, \{(\alpha_1 \alpha_2 - \omega^2) - \omega^2 \varphi'(\alpha_1 + \alpha_2)\}\,,$$
$$\eta = \omega\{\beta(\varphi' - \varphi) - \alpha_2 \varphi\}\, (\alpha_1 + \alpha_2) + \omega\varphi(\alpha_1 \alpha_2 - \omega^2)\,,$$
$$\eta' = \omega\{\gamma(\varphi - \varphi') - \alpha_2 \varphi'\}\, (\alpha_1 + \alpha_2) + \omega(\alpha_1 \alpha_2 - \omega^2)\,. \tag{17}$$

(iii) Pressure gradient varies exponentially with time. Suppose

$$f(t) = \sum_{r=0}^{\infty} a_r\, e^{-rt}\,, \qquad a_r \text{ are constants}\,.$$

The expressions for the velocity of the two phases in this case are obtained as

$$u = \frac{4}{h} \sum_{n=0}^{\infty} \sum_{m=0}^{\infty} \frac{(-1)^{n+m}}{q_n q_m} \left\{ \sum_{r=0}^{\infty} a_r \left[\frac{[\beta\varphi' - (\alpha_2 + \beta + \gamma)\, \varphi]\, e^{-rt}}{(\alpha_1 + \gamma)\, (\alpha_2 + \gamma)} + \right. \right.$$
$$\left. \left. + \frac{1}{\beta + \gamma} \left\{ \frac{(\beta\varphi' + \gamma\varphi)\, e^{\alpha_1 t}}{\alpha_1 + \gamma} + \frac{\beta(\varphi - \varphi')\, e^{\alpha_2 t}}{\alpha_2 + \gamma} \right\} \right] \right\} \cos q_n x \cos q_m y \tag{18}$$

and

$$v = \frac{4}{h} \sum_{n=0}^{\infty} \sum_{m=0}^{\infty} \frac{(-1)^{n+m}}{q_n q_m} \left\{ \sum_{r=0}^{\infty} a_r \left[\frac{[\gamma\varphi - (\alpha_2 + \gamma + \beta)\, \varphi']\, e^{-rt}}{(\alpha_1 + \gamma)\, (\alpha_2 + \gamma)} + \right. \right.$$
$$\left. \left. + \frac{1}{\beta + \gamma} \left\{ \frac{(\gamma\varphi + \beta\varphi')\, e^{\gamma_1 t}}{\alpha_1 + \gamma} + \frac{\gamma(\varphi' - \varphi)\, e^{\alpha_2 t}}{\alpha_2 + \gamma} \right\} \right] \right\} \cos q_n x \cos q_m y\,. \tag{19}$$

5. Numerical Results and Discussions

In order to gain some insight into the patterns of flow, velocity, profiles for fluid phase (figs. 1–4) and particulate phase (fig. 5–6) have been plotted for different values of variables. For numerical evaluation of the results we have taken the fluid density ϱ to be 1.16 g/cm³, $\bar{\varrho}_p$, the density of the material of the particulate phase to be 11.34 g/cm³, and the volume fraction of the particulate phase $\varrho_p/\bar{\varrho}_p = 0.1$. The numerical studies reveal that the fluid phase velocity is greater than the corresponding particulate phase velocity. The effect of increasing the parameter γ is to decrease the fluid phase velocity and to increase the particulate phase velocity. Further, the parameter ω has more influence on the fluid phase velocity than the particulate phase.

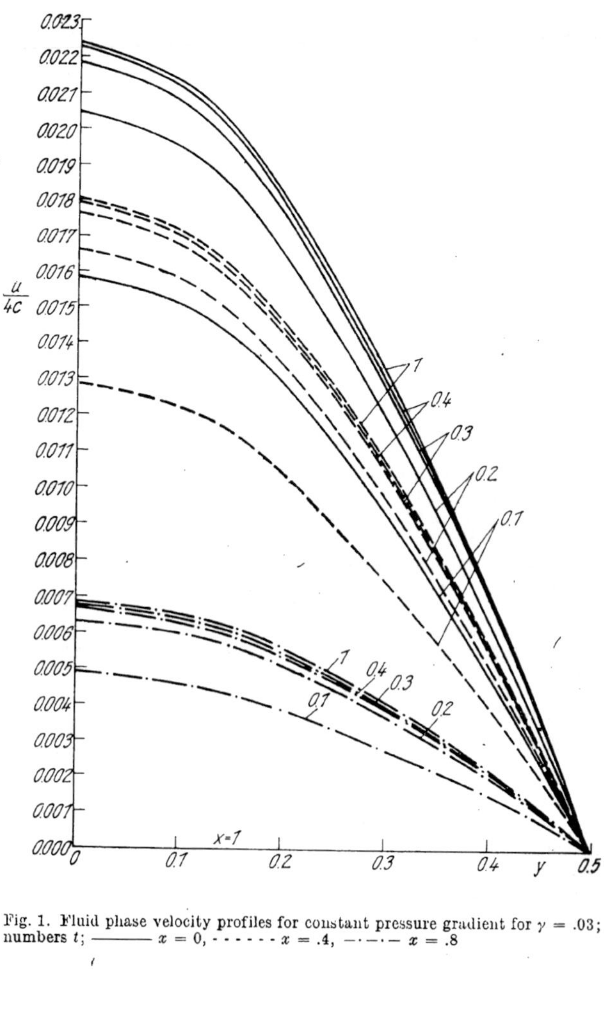

Fig. 1. Fluid phase velocity profiles for constant pressure gradient for $\gamma = .03$; numbers t; ——— $x = 0$, - - - - - - $x = .4$, —·—·— $x = .8$

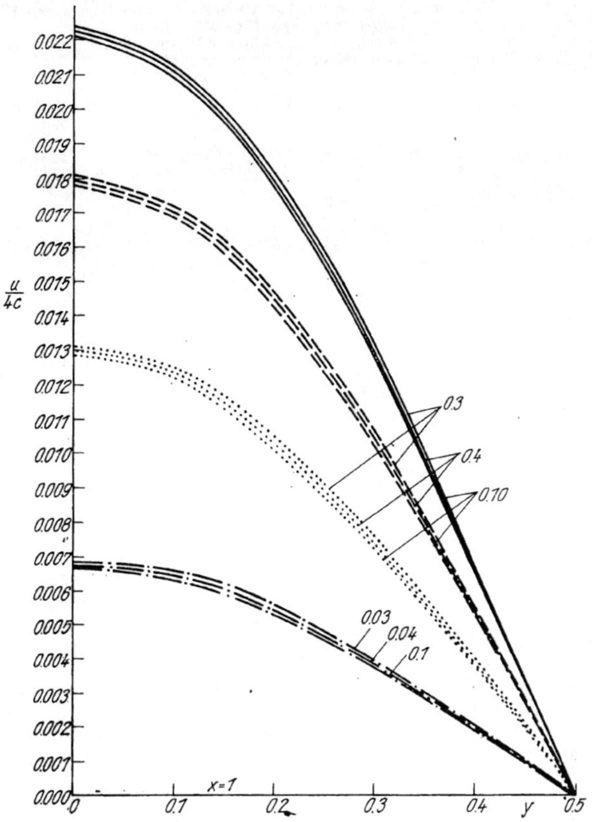

Fig. 2. Fluid phase velocity profiles for constant pressure gradient for $t = 1$; numbers γ; ——— $x = 0$, - - - - - - $x = .4$, · · · · · · $x = .6$, —·—·— $x = .8$.

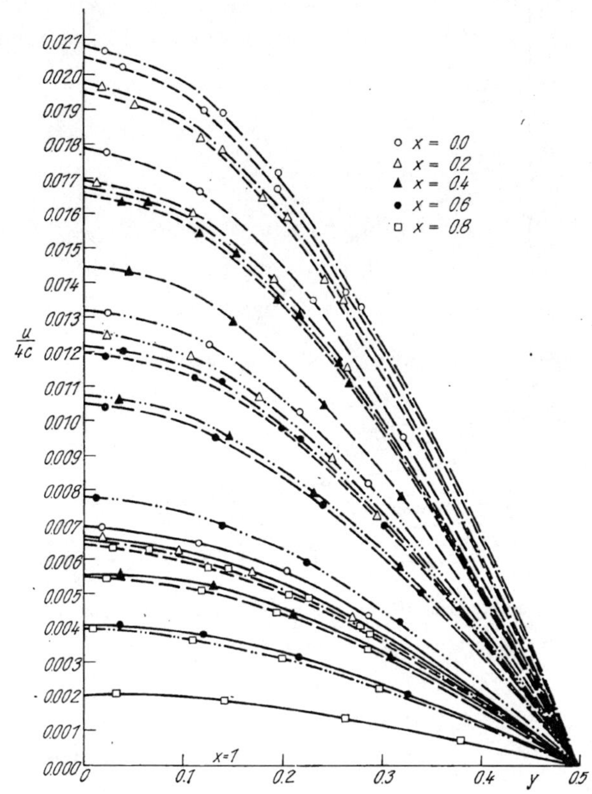

Fig. 4. Fluid phase velocity profiles for sinusoidal pressure gradient for $t = .4$; $\nu = .03$; ——— $\omega = 1$, —··—··— $\omega = 2$, — — — $\omega = 3$, - - - - - - $\omega = 4$, — —·— $\omega = 5$

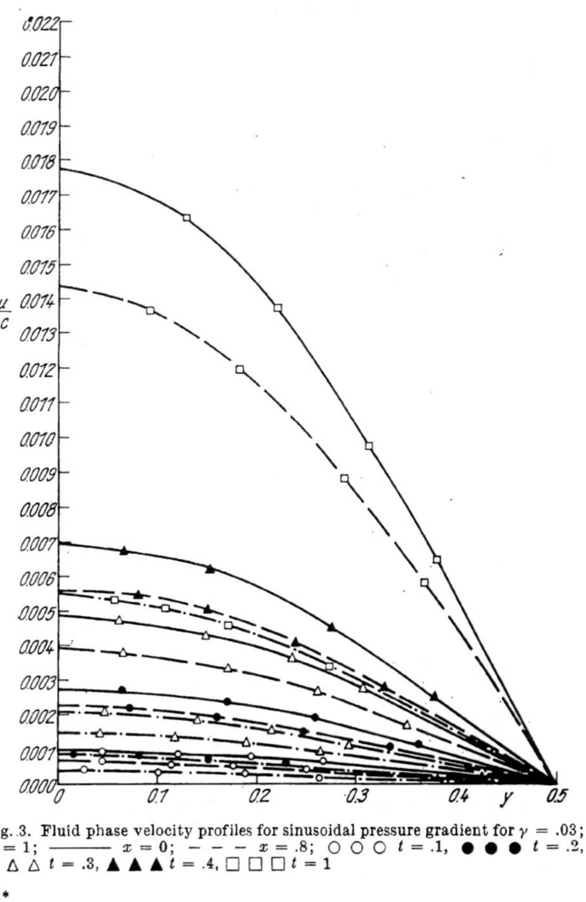

Fig. 3. Fluid phase velocity profiles for sinusoidal pressure gradient for $\gamma = .03$; $\omega = 1$; ——— $x = 0$; — — — $x = .8$; ○ ○ ○ $t = .1$, ● ● ● $t = .2$, △ △ △ $t = .3$, ▲ ▲ ▲ $t = .4$, □ □ □ $t = 1$

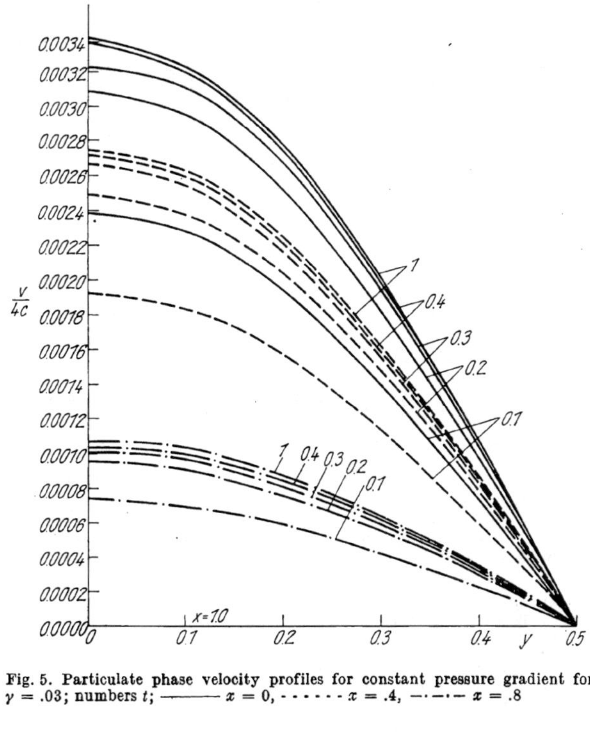

Fig. 5. Particulate phase velocity profiles for constant pressure gradient for $\gamma = .03$; numbers t; ——— $x = 0$, - - - - - - $x = .4$, —·—·— $x = .8$

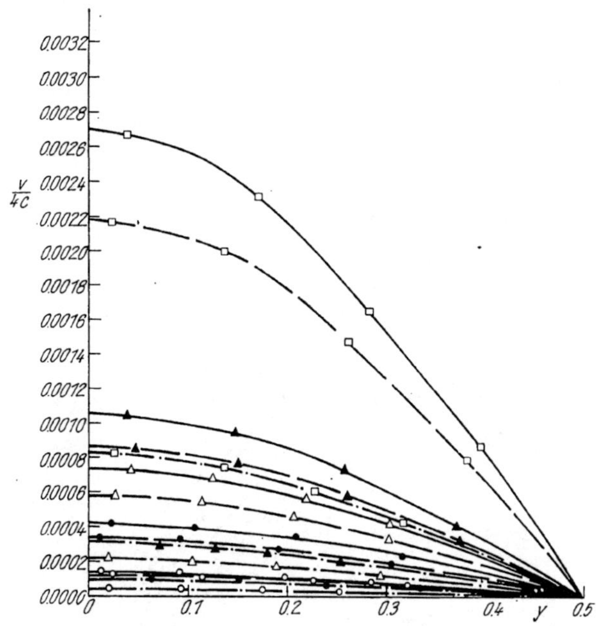

Fig. 6. Particulate phase velocity profiles for sinuoidal pressure gradient for $\gamma = .03$; $\omega = 1$; ——— $x = 0$, — — — $x = .4$, —·—·— $x = .8$, ○ ○ ○ $t = .1$, ● ● ● $t = .2$, △ △ △ $t = .3$, ▲ ▲ ▲ $t = .4$, □ □ □ $t = .5$

Acknowledgement

The work was supported by a Project-grant from the Ministry of Education and Social Welfare, Govt. of India.

References

1 TRUESDELL, C., Atti Accad. Naz. Lincei RC. Ser 8, vol. **22**, pp. 33–158, 1957 (Also translation by the author in continuum Mechanics II. The Rational Mechanics of Materials, pp. 295, 1965, (Ed. TRUESDELL, C.) Gordon and Breach, New York).

2 TRUESDELL, C.; TOUPIN, R., The Classical Field Theories, Handbuch der Physik (Ed. S. FLÜGGE) III/I pp. 226 (Springer-Verlag, Berlin, 1960).

3 ERINGEN, A. C.; INGRAM, J. D., A Continuum Theory of Chemically Reacting Media-I, International Journal of Engineering Science, vol. **3**, pp. 197–212 (1965).

4 GREEN, A. E.; NAGHDI, P. M., A Dynamical Theory of Interacting Continua, International Journal of Engineering Science, vol. **3**, pp. 231–241 (1965).

5 MILLS, N., Incompressible Mixture of Newtonian Fluids, International Journal of Engineering Science, Vol. **4**, pp. 97–112, (1966).

6 MILLS, N., On a Theory of Multi-component Mixture, Quarterly Journal of Mechanics and Applied Mathematics, vol. **20**, pp. 499–508 (1967).

7 ATKIN, R. J.; CRAINE, R. E., Continuum Theories of Mixture: Basic Theory and Historical Development, Quarterly Journal of Mechanics and Applied Mathematics, vol. **29**, pp. 209–244 (1976).

8 ATKIN, R. J.; CRAINE, R. E., Continuum Theories of Mixtures: Application, Journal of Institute of Mathematics and Application, vol. **17**, pp. 153–207 (1976).

9 WILHELM, H. E.; VAN DER WERFF, T. J., Oscillating Flows of Mixed Fluids, Applied Scientific Research, vol. **33**, pp. 339–352 (1977).

10 MARBLE, F. E., Dynamics of a Gas Containing Small Solid Particles, Presented at the Fifth AGARD Combustion and Propulsion Colliquium, Braunschweig (1962).

11 MARBLE, F. E., Dynamics of Dusty Gas, Annual Review of Fluid Mechanics, Vol. **2**, pp. 397–446 (1971).

12 SOO, S. L., Fluid Dynamics of Multiphase Systems, Blaisdell Publishing Company, Waltham, Mass (1967).

13 DAVIDSON, J. F.; HARRISON, D., Fluidized Particles, Cambridge Press, London (1963).

14 MURRAY, J. D., On the Mathematics of Fluidization, Part 1: Fundamental Equations and Wave Propagation, Journal of Fluid Mechanics, Vol. **21**, pp. 465–493 (1965).

15 MURRAY, J. D., Technical Report 1, Grant AP00455-01, National Centre for Pollution Control, U.S. Public Health Service (1966).

16 MURRAY, J. D., Some Basic Aspects of One-Dimensional Incompressible Particle-Fluid Two-phase Flows, Astronautica Acta, vol. **13**, pp. 417–430 (1967).

17 BRENNER, H., Rheology of Two-phase Systems, Annual Review of Fluid Mechanics, Vol. **2**, pp. 131–176 (1970).

18 DIGIOVANNI, P. R.; LEE, S. L., Impulsive Motion in a Particle-Fluid Suspension Including Particulate Volume, Density and Migration Effects, Trans ASME, Journal of Applied Mechanics, Vol. **41**, Number 1, pp. 35–41 (1974).

19 DREW, D. A.; SEGEL, L. A., Averaged Equations for Two-phase Flows, Studies in Applied Mathematics, vol. **50**, pp. 205–231 (1971).

20 DREW, D. A., Low Concentration Two-phase Flow Near a Stagnation Point, Physics of Fluids, vol. **17**, pp. 1685–1691 (1974).

21 DREW, D. A., Macroscopic Streamline Integral Relations for Two-phase Flows, Trans. ASME, Journal of Applied Mechanics, vol. **42**, pp. 766–770 (1975).

22 DREW, D. A., Two Phase Flows: Constitutive Equations for Lift and Brownian Motion and Some Basic Flows, Arch. Rational Mechanics and Analysis, vol. **62**, pp. 149–163 (1976).

23 DREW, D. A., Lubrication Flow of a Particle Fluid Mixture, Trans. ASME, Journal of Applied Mechanics, vol. **6**, pp. 221–213 (1979).

24 PEDDISON, J., Some unsteady Parallel Flows of Particulate Suspensions, Applied Scientific Research, vol. **32**, pp. 239–268 (1976).

Eingereicht am 17. 8. 1980

Anschrift: Dr. R. P. AGARWAL, Department of Mathematics, M.N.R. ENGG. COLLEGE, Allahabad, India

ZAMM 62, 264–265 (1982)

T. NAKAGAWA

The Comparison of a New Theory of Grid-produced Turbulence with an Experiment

1. Introduction

In a previous paper by the author [1], grid-produced turbulence has been solved as an initial value problem and the turbulent energy decay law has been derived under the assumptions that the turbulence behind the grid is isotropic and the pressure-velocity correlations are negligibly small. It has been found that the grid-produced turbulence decays depending on the distance from the grid and the Reynolds number based on the mean flow velocity and the characteristic length of the grid as well as the initial conditions including effects such as grid geometry and porosity. Contrary to the classical theory [2], it is of importance to note that the new theory has no empirical constants in the turbulent energy decay law.

In the present paper, the author's theory [1] has been compared with BATCHELOR's and TOWNSEND's [2] experiment. It is found that the theory is applicable for a limited range $300 \lesssim \hat{x} \lesssim 800$, where $\hat{x}$ is the distance from the grid normalized by the grid mesh size.

2. The New Turbulent Energy Decay Law

When turbulence is considered to be isotropic, the turbulent energy ε per unit mass of fluid may be expressed by

$$\varepsilon = \tfrac{3}{2}\overline{u'^2}\,. \tag{1}$$

The decay of $\overline{u'^2}$ will be, therefore, discussed herein rather than that of ε.

The present author [1] has derived a turbulent energy decay law using a simplified turbulent model in which the isotropic turbulence is linearly superimposed on the uniform mean flow,

$$\vec{u} = (U, 0, 0)\,. \tag{2}$$

The turbulent energy decay law has been expressed by

$$\overline{\hat{u}'^2} = \overline{\hat{u}'^2_0}\left[1 + \frac{8(\hat{x} - \hat{x}_0)}{a^2 R_M}\right]^{-5/2} \tag{3}$$

where $\overline{\hat{u}'^2} = \overline{u'^2}/U^2$, $\overline{\hat{u}'^2_0} = \overline{u'^2_0}/U^2$, $\hat{x} = x/M$, $\hat{x}_0 = x_0/M$ and $R_M = MU/\nu$, the subscript [0] denotes a position where the longitudinal double velocity correlation function f attains a Gaussian distribution, x the distance from the grid, M the grid mesh size and a is a constant, to be obtained from the longitudinal double velocity correlation function

$$f = \exp(-\hat{r}^2/a^2)\,, \tag{4}$$

at $\hat{x} = \hat{x}_0$, where $\hat{r} = r/M$ is the normalized distance between the two points. It must be noted that the longitudinal double velocity correlation function f at this point is nothing else but the initial condition in the author's theory [1].

3. The Comparison of the Theory with the Data

Turbulent energy decay behind the grid and the longitudinal double velocity correlation function f at $\hat{x} = 320$ have been measured by BATCHELOR and TOWNSEND [2]. Before the actual comparison of the theory with the data, it is necessary to find the constant value a in Eq. (3). Figure 1 shows the longitudinal double velocity correlation function measured by BATCHELOR

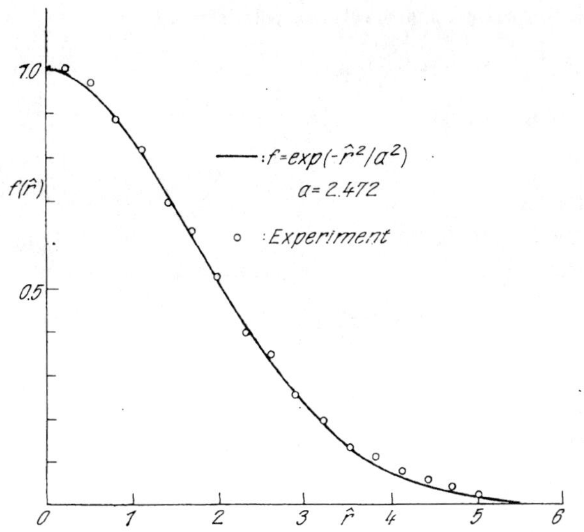

Fig. 1. Longitudinal double velocity correlation function at $\hat{x} = 320$. $M = 0.159$ cm, $U = 620$ cm/s, $R_M = 650$

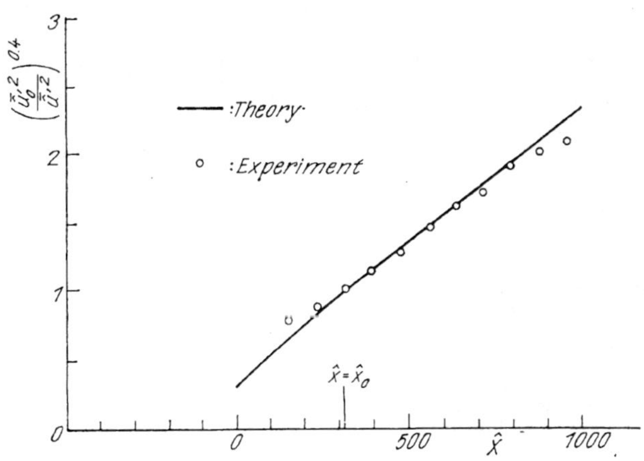

Fig. 2. Decay of grid-produced turbulent energy. For legend cf. Figure 1

and TOWNSEND [2] and the best fitted Gaussian distribution from which the value of the constant $a = 2.472$ has been obtained. Substituting $a = 2.472$ and $R_M = 650$ into Eq. (3), it is now possible to compare the turbulent energy decay law with the data. The turbulent energy at $\hat{x} = \hat{x}_0$ is, however, equal to $\overline{\hat{u}'^2_0}$, so that it may be more convenient to rewrite Eq. (3) into the form

$$(\overline{\hat{u}'^2_0}/\overline{\hat{u}'^2})^{0.4} = 1 + \frac{8(\hat{x} - \hat{x}_0)}{a^2 R_M}. \tag{5}$$

Figure 2 provides the comparison between the above turbulent energy decay law and BATCHELOR's and TOWNSEND's data. It may be evident that the author's theory of grid-produced turbulence agrees with the data for a limited range $300 \lesssim \hat{x} \lesssim 800$ of the distance behind the grid, but deviates from the data as aparting from both ends of that range.

4. Discussions and Conclusions

The dispersion relationships for the present theory of grid-produced turbulence have been discussed by TSUGÉ [3] and the present author [4] incorporating the pressure-velocity correlations that have been neglected in the author's previous paper [1]. These discussions regarding the dispersion relationships have shown that the grid-produced turbulence consists of the OSEEN mode solution and the LAPLACE mode solution. It has also become clear that the author's solution in the previous paper [1] is nothing but the OSEEN mode solution, while the LAPLACE mode solution has been lost by the tentative negligence of the pressure-velocity correlations. Hence, it may be considered that the partial disagreements between the author's theory [1] and BATCHELOR's and TOWNSEND's [2] experiment are mainly caused by the neglected LAPLACE mode solution though the closure assumption (or negligence of the triple velocity correlations) and the isotropic assumption of turbulence may also introduce possible errors into the theoretical results. Nevertheless, the present comparison shows that the theory [1] is applicable for the range $300 \lesssim \hat{x} \lesssim 800$ of the distance behind the grid.

Finally, it may be worth pointing out the clarified physical character of the grid-produced turbulence by means of the consideration of the dispersion relationships: The OSEEN mode of the grid-produced turbulence propagates with the same speed as the mean flow, whereas the LAPLACE mode is stationary relative to the grid. Moreover, the LAPLACE mode decays more rapidly than the OSEEN mode.

References

1 NAKAGAWA, T., A theory on decay of grid produced turbulence, ZAMM 59 (1979) 648–651.
2 BATCHELOR, G. K.; TOWNSEND, A. A., Decay of turbulence in the final period, Proc. Roy. Soc. London, A 194 (1948) 527–543.
3 TSUGÉ, S., Methods of separation of variables in turbulence theory, NASA Contractor Report 3054, Oct. 1978.
4 NAKAGAWA, T., Some recent progress in the theory of grid-produced turbulence and its significance in wind engineering, Proc. of 6th National Symposium on Wind Engineering, Tokyo, Nov. 1980, 59–66 (in Japanese).

Eingereicht am 27. 12. 1980, in revidierter Fassung am 8. April 1981

Anschrift: Dr. TAKEO NAKAGAWA, associate Professor, Department of Civil Engineering, Kanazawa Institute of Technology, Nonoichi Kanazawa 921, Japan

ZAMM 62, 265–267 (1982)

A. PRECHTL

On the Interaction of Electric Fields in Non-Conducting Elastic Fluids

A perturbation procedure is employed to derive tractable equations describing the interaction of inhomogeneous electric fields in elastic polarizable fluids. The electrical generation of sound waves is discussed by example.

1. Introduction

The purpose of this paper is to derive equations which can be treated analytically describing the interaction of quasi-electrostatic fields in polarizable elastic fluids. Although the media are assumed to be perfect isolators such interactions occur because of the electrostrictive effect and the body force exerted on the dipole distribution, on the one hand and the density dependence of the polarizability on the other.

The resulting effects may be expected to be of some importance in connection with strongly inhomogeneous electric fields. Hence, a perturbation procedure is employed according to which the undisturbed quasi-electrostatic field problem has to be solved first. Small deviations from the uniform density are then governed by a wave equation with a space and time varying driving term thus leading to distributed sources of the acoustic field.

2. The basic equations

In a lossless polarizable fluid medium under quasi-electrostatic conditions the macroscopic velocity field $\boldsymbol{u}$ with respect to an inertial frame and the hydrostatic pressure p, have to satisfy the equation of motion [1]

$$\frac{1}{v}\frac{d\boldsymbol{u}}{dt} = -\nabla p + \boldsymbol{P}\cdot\nabla\boldsymbol{E}, \tag{2.1}$$

where v is the specific volume (inverse mass density) and

$$\frac{d}{dt} = \frac{\partial}{\partial t} + \boldsymbol{u}\cdot\nabla \tag{2.2}$$

denotes the material time derivative. The electric polarization

$\boldsymbol{P}$ and the electric field strength $\boldsymbol{E}$ are restricted by the quasi-electrostatic equations

$$\nabla \times \boldsymbol{E} = \boldsymbol{0}, \qquad \nabla \cdot (\varepsilon_0 \boldsymbol{E} + \boldsymbol{P}) = 0. \tag{2.3}$$

In addition, the conservation of mass requires

$$\frac{dv}{dt} = v \nabla \cdot \boldsymbol{u}, \tag{2.4}$$

while the conservation of energy ensures the existence of a state function

$$w = w(v, E^2), \tag{2.5}$$

in terms of which the pressure p and the susceptibility per unit mass χ_m are given by

$$p = -\frac{\partial w}{\partial v}, \qquad \chi_m = -\frac{2}{\varepsilon_0}\frac{\partial w}{\partial E^2}, \tag{2.6}$$

as functions of v and E^2. Thus we have the general constitutive equations

$$p = p(v, E^2), \qquad \boldsymbol{P} = \frac{1}{v}\varepsilon_0 \chi_m(v, E^2)\, \boldsymbol{E} \tag{2.7}$$

where the functions p and χ_m satisfy the relation

$$\frac{\partial p}{\partial E^2} = \frac{\varepsilon_0}{2}\frac{\partial \chi_m}{\partial v} \tag{2.8}$$

valid for isentropic processes. Supplemented by appropriate initial and boundary conditions the set of equations (2.1)—(2.4) and (2.7) provides a complete description.

3. Perturbations of a uniform reference state

Let v_0 denote the constant specific volume of a reference state in which the fluid is at rest and the electric field strength vanishes. We first expand the state function (2.5) in powers of the relative variation

$$\delta = (v - v_0)/v_0 \tag{3.1}$$

and of E^2 up to the second order,

$$w(v, E^2) = w_0 - p_0 v_0 \left(1 - \frac{\varkappa}{2}\delta\right)\delta - \frac{\varepsilon_0}{2}\chi_0 v_0 \left(1 + \zeta\delta + \frac{\nu}{2}E^2\right)E^2, \tag{3.2}$$

where the coefficients, w_0,

$$p_0 = -\left(\frac{\partial w}{\partial v}\right)_0, \qquad \varkappa = \frac{v_0}{p_0}\left(\frac{\partial^2 w}{\partial v^2}\right)_0, \qquad \chi_0 = -\frac{2}{\varepsilon_0 v_0}\left(\frac{\partial w}{\partial E^2}\right)_0,$$

$$\nu = \left(\frac{\partial^2 w}{\partial E^4}\Big/\frac{\partial w}{\partial E^2}\right)_0, \qquad \zeta = v_0\left(\frac{\partial^2 w}{\partial v\, \partial E^2}\Big/\frac{\partial w}{\partial E^2}\right)_0 \tag{3.3}$$

are evaluated at $v = v_0$ and $E^2 = 0$. p_0, $\varkappa p_0$ and χ_0 denote the pressure, the adiabatic module of elasticity and the linear susceptibility of the reference state, respectively, while ζ is the electrostrictive coefficient and ν accounts for the electric nonlinearity in a first approximation. Then, from (2.6), the expressions

$$p = p_0\left(1 - \varkappa\delta + \zeta\frac{\varepsilon_0\chi_0}{2p_0}E^2\right), \qquad \chi_m = v_0\chi_0(1 + \nu E^2 + \zeta\delta) \tag{3.4}$$

are obtained.

Let $\boldsymbol{E}_s$ be the electric field strength as solution of the equations (2.3) with $\delta = 0$, i.e.,

$$\nabla \times \boldsymbol{E}_s = \boldsymbol{0}, \qquad \nabla \cdot (\varepsilon_0 \boldsymbol{E}_s + \boldsymbol{P}_s) = 0, \tag{3.5}$$

with

$$\boldsymbol{P}_s = \varepsilon_0\chi_0(1 + \nu E_s^2)\, \boldsymbol{E}_s. \tag{3.6}$$

If E_0 designates a constant characteristic magnitude of $\boldsymbol{E}_s$ (e.g., $E_0 = \max|\boldsymbol{E}_s|$) the perturbed electric field is taken as

$$\boldsymbol{E} = \boldsymbol{E}_s + E_0\boldsymbol{e}, \qquad |\boldsymbol{e}| \ll 1. \tag{3.7}$$

The values of the parameters

$$n = \nu E_0^2, \qquad b = \frac{\varepsilon_0\chi_0}{2p_0}E_0^2 \tag{3.8}$$

are usually very small so that the products with the perturbations as well as nonlinear terms in the perturbations can be neglected if compared with unity. From (3.4) and $(2.7)_2$ we then have the pressure and the electric polarization

$$p = p_0\left[1 - \varkappa\delta + b\zeta\left(\frac{E_s}{E_0}\right)^2\right],$$
$$\boldsymbol{P} = \boldsymbol{P}_s + \varepsilon_0\chi_0 E_0\left[\boldsymbol{e} - (1 - \zeta)\frac{\boldsymbol{E}_s}{E_0}\delta\right], \tag{3.9}$$

while the force density in the equation of motion (2.1) may be written as

$$\boldsymbol{P} \cdot \nabla \boldsymbol{E} = p_0 b \nabla (E_s/E_0)^2. \tag{3.10}$$

The linearized form of the latter equation becomes therefore

$$\frac{\partial \boldsymbol{u}}{\partial t} = c_s^2 \nabla\left[\delta + \frac{b}{\varkappa}(1 - \zeta)\left(\frac{E_s}{E_0}\right)^2\right] \tag{3.11}$$

with

$$c_s = \sqrt{v_0 p_0 \varkappa}, \tag{3.12}$$

and the continuity equation (2.4) renders

$$\frac{\partial \delta}{\partial t} = \nabla \cdot \boldsymbol{u}. \tag{3.13}$$

Taking the relations (3.5) for the undisturbed electric field into account we further obtain from (2.3) and (3.7) with $(3.9)_2$

$$\nabla \times \boldsymbol{e} = \boldsymbol{0}, \qquad \nabla \cdot \left[\boldsymbol{e} - \frac{\chi_0}{1 + \chi_0}(1 - \zeta)\frac{\boldsymbol{E}_s}{E_0}\delta\right] = 0. \tag{3.14}$$

In summarizing, first the electrostatic field problem (3.5) has to be solved. Inserting the solution $\boldsymbol{E}_s$ into (3.11) this equation together with (3.13) results in an inhomogeneous wave equation for the relative variation δ of the specific volume. In fact, since E_s^2 is multiplied by the small parameter b, only the solution of the linear problem (3.5) with

$$\boldsymbol{P}_s = \varepsilon_0\chi_0\boldsymbol{E}_s, \tag{3.15}$$

instead of (3.6) is required. We note that the propagation of the density disturbance is not influenced by a homogeneous electric field in our approximation. Finally, the perturbation $\boldsymbol{e}$ of the electric field is found from (3.14).

4. Electrically generated sound waves

To demonstrate the applicability of the equations derived we consider the following example. A rigid metallic sphere of radius R with prescribed surface charge density

$$\sigma(t) = \sigma_0 s(t) \tag{4.1}$$

is situated in a fluid medium of infinite extent. The perturbations of the specific volume and of the electric field are to be determined.

Because of the spherical symmetry the electric field strength $\boldsymbol{E}_s$ as solution of the linear field problem (3.5) with (3.15) is radially directed and can be written down immediately as function of the radial coordinate r and time t,

$$E_s = E_0 s(t)\,(R/r)^2. \tag{4.2}$$

Here, the reference value E_0 has been chosen as

$$E_0 = \frac{\sigma_0}{\varepsilon_0(1 + \chi_0)}. \tag{4.3}$$

Taking the small nonlinear term in the constitutive relation (3.6) into account the corrected values of E_s are

$$E_s = E_0\left[1 - \frac{n\chi_0}{1 + \chi_0}s^2(t)\left(\frac{R}{r}\right)^4\right]s(t)\left(\frac{R}{r}\right)^2. \tag{4.4}$$

Direct integration of (3.14) relates the perturbations by

$$e = \frac{\chi_0}{1 + \chi_0}(1 - \zeta)\, s(t)\left(\frac{R}{r}\right)^2\delta, \tag{4.5}$$

if e denotes the radial projection of $\boldsymbol{e}$. It remains to determine δ by means of the inhomogeneous wave equation

$$\frac{1}{c_s^2}\frac{\partial^2 \delta}{\partial t^2} - \Delta\delta = \frac{b}{\varkappa}(1 - \zeta)\, s^2(t)\, \Delta\left(\frac{R}{r}\right)^4, \tag{4.6}$$

with the spherical Laplacian

$$\Delta = \frac{1}{r^2}\frac{\partial}{\partial r}\left(r^2\frac{\partial}{\partial r}\right), \tag{4.7}$$

subject to the boundary conditions

$$r = R: \quad \frac{\partial \delta}{\partial r} = 4 \frac{b}{\varkappa} (1 - \zeta) s^2(t), \qquad r \to \infty: \quad \delta \to 0. \tag{4.8}$$

The latter follow from (3.11) together with the requirement $\boldsymbol{u}(R, t) = \boldsymbol{0}$, and $\boldsymbol{v} \to \boldsymbol{v}_0$ as $r \to \infty$.

To simplify expressions we introduce the normalized independent variables

$$\tau = \frac{c_s}{R} t, \quad \xi = \frac{r}{R}, \tag{4.9}$$

and change the dependent variable according to

$$\psi(\xi, \tau) = \frac{r}{R} \delta(r, t). \tag{4.10}$$

This reduces equation (4.6) to the inhomogeneous plane wave equation

$$\frac{\partial^2 \psi}{\partial \tau^2} - \frac{\partial^2 \psi}{\partial \xi^2} = 12a \frac{S(\tau)}{\xi^5}, \tag{4.11}$$

with

$$a = \frac{b}{\varkappa} (1 - \zeta), \qquad S(\tau) = s^2(t), \tag{4.12}$$

and the boundary conditions

$$\xi = 1: \quad \frac{\partial \psi}{\partial \xi} - \psi = 4aS(\tau), \qquad \xi \to \infty: \quad \frac{\psi}{\xi} \to 0. \tag{4.13}$$

The static solution $\psi = -a/\xi^3$ is obtained by setting $\partial/\partial \tau = 0$ and $S = 1$. In the original variables and with (4.4) we have in this case

$$v = v_0 \left[1 - a \left(\frac{R}{r}\right)^4\right], \qquad E = E_0 \left(\frac{R}{r}\right)^2 - \frac{\chi_0}{1 + \chi_0} E_0[n + (1 - \zeta) a] \left(\frac{R}{r}\right)^6. \tag{4.14}$$

Suppose now the excitation is harmonic,

$$s(t) = \cos \omega t, \tag{4.15}$$

i.e.,

$$S(\tau) = \frac{1}{2} (1 - \cos \alpha \tau), \qquad \alpha = 2 \frac{\omega R}{c_s}. \tag{4.16}$$

A particular solution of (4.11) may be assumed in the form

$$\psi_p = -a \left[\frac{1}{2\xi^3} + \Phi(\xi) \cos \alpha \tau\right]. \tag{4.17}$$

The first term is suggested by the static solution while the second represents a standing wave. This leads to the ordinary differential equation

$$\Phi''(\xi) + \alpha^2 \Phi(\xi) + 6/\xi^5 = 0 \tag{4.18}$$

with general solution

$$\Phi(\xi) = -\frac{1}{2\xi^3} + \frac{\alpha^2}{4\xi} [1 - \alpha \xi f(\alpha \xi)] + C_1 \cos \alpha \xi + C_2 \sin \alpha \xi. \tag{4.19}$$

The auxiliary function f has the integral representation

$$f(x) = \int_0^\infty \frac{\sin t}{t + x} \, dt, \qquad x > 0, \tag{4.20}$$

and is tabulated in [2, number 5.2.]. Superimposing the general solution of the homogeneous part of (4.11) the result can be matched to the boundary conditions (4.13) without difficulties. We obtain

$$v = v_0(1 + \delta),$$

$$\frac{1}{a} \delta = -\frac{1}{\xi^4} \left(\cos \frac{\alpha}{2} \tau\right)^2 - \frac{\alpha^2}{4\xi^2} [1 - \alpha \xi f(\alpha \xi)] \cos \alpha \tau - \frac{\alpha^3}{4\xi} \frac{\pi}{2} \cos \alpha \xi \cos \alpha \tau + A(\alpha) \frac{1}{\xi} \cos [\alpha(\tau - \xi + 1) - \beta], \tag{4.21}$$

where

$$A(\alpha) = \frac{\alpha^3}{4} \frac{\pi}{2} \cos (\alpha - \beta) + \frac{\alpha^2}{4} (1 + \alpha^2)^{-1/2} [2 - \alpha f(\alpha) + \alpha^2 f'(\alpha)], \qquad 0 \leqq \beta = \arctan \alpha < \frac{\pi}{2}. \tag{4.22}$$

In the form (4.21) the relative disturbance of the specific volume is composed of the following parts. The first term is the static solution multiplied by the excitation function $S(\tau)$ while the second term represents a standing wave whose amplitude has the asymptotic form $a/(2\xi^4)$. The third term is again a standing wave and the fourth an outgoing harmonic spherical wave. The amplitudes of all components are proportional to the parameter a and are partly dependent on the relative frequency α. In the limit $\alpha \gg 1$, i.e., $\omega \gg c_s/(2R)$, we have

$$A(\alpha) \approx \frac{\alpha^3}{4} \frac{\pi}{2} \sin \alpha \tag{4.23}$$

and (4.21) reduces to

$$\frac{1}{a} \delta \approx -\frac{\alpha^3}{4\xi} \frac{\pi}{2} [\cos \alpha \xi \cos \alpha \tau - \sin \alpha \sin \alpha(\tau - \xi + 1)]. \tag{4.24}$$

On the other hand, for small relative frequencies, $\alpha \ll 1$, the function A is approximated by $A(\alpha) \approx \alpha^2/2$ resulting in

$$\frac{1}{a} \delta \approx -\frac{1}{\xi^4} \left(\cos \frac{\alpha}{2} \tau\right)^2 \tag{4.25}$$

in the vicinity $\xi \gtrsim 1$ of the sphere. This is the modulated static solution.

References

1 Tiersten, H. F., On the nonlinear equations of thermo-electroelasticity, Int. J. Engng. Sci. 9 (1971) 587–604.

2 Abramovitz, M.; Stegun, I. A., (ed.): Handbook of Mathematical Functions, 8th printing, New York: Dover.

Eingereicht am 23. 6. 1981

Anschrift: Dr. A. Prechtl, Institute for Electrical Machines, Technical University of Vienna, Gusshausstraße 25, A-1040 Vienna, Austria

ZAMM 62, 267–269 (1982)

B. Gampert, P. Wagner

Der Einfluß der Molmasse polymerer Additive auf die Reibungsverminderung

Einleitung

Gibt man der turbulenten Rohrströmung einer Flüssigkeit Makromoleküle bei, so verringert sich bei konstantem Durchfluß der Druckabfall um bis zu 80%. Von den wasserlöslichen Polymeren haben sich Polyäthylenoxyd (PEO) und Polyacrylamid (PAA) als besonders wirksam erwiesen. In starken Scherfeldern neigen die Makromoleküle allerdings dazu, in kleinere, weniger wirksame Moleküle zu zerfallen. Erklärungsversuche dieser nach ihrem Entdecker auch als Toms-Effekt bezeichneten Widerstandsverminderung sind erschwert durch die Streuung der in der Literatur angegebenen experimentellen Daten und deren schlechte Reproduzierbarkeit.

Die unzureichende Charakterisierung der Experimente betrifft insbesondere die wichtigen Parameter der Molmasse und der Molmassenverteilung. Die im allgemeinen benutzten technischen Substanzen weisen ein breites und zufälliges Molmassenspektrum auf. Die mittlere Molmasse ist aber nur dann als Parameter aussagekräftig, wenn die Molmassenverteilung hinreichend eng ist.

Die vorliegende Arbeit untersucht den Einfluß der Molmasse in Wasser gelösten Polyacrylamids auf Druckabfall und Degradation einer turbulenten, ausgebildeten Rohrströmung. Um die Schwierigkeiten der präparativen Fraktionierung zu vermeiden [1], wurde hochreines Polyacrylamid (PAA) im Labor selbst hergestellt und die Polymerisation so geführt, daß hinreichend enge Molmassenfraktionen entstanden.

Versuchsdurchführung

Polyacrylamid (PAA) wurde je nach gewünschter Molmasse in einer wäßrigen Lösung bidestillierten Wassers oder einer Wasser-Methanol-Lösung unter N_2-Atmosphäre radikalisch polymerisiert. Als Initiator diente H_2O_2, da dieses im Gegensatz zu $K_2S_2O_8$, das bei der technischen Herstellung im allgemeinen verwendet wird, nach der Polymerisation zu Wasser zerfällt. Das PAA wurde aus der wäßrigen Lösung in reinem Methanol bei Zimmertemperatur ausgefällt, nach Entfernung der Methanolreste in bidestilliertem Wasser gelöst und gefriergetrocknet. Die Polymerisationstemperatur lag zwischen 35 °C und 70 °C.

Die mittlere Molmasse (Viskositätsmittel) der in dieser Arbeit untersuchten PAA-Fraktion beträgt $\bar{M}_\eta = 3{,}24 \cdot 10^6$ g/mol für die Probe a, $\bar{M}_\eta = 4{,}6 \cdot 10^6$ g/mol für die Probe b und $5{,}03 \times 10^6$ g/mol für die Probe c.

Für wäßrige PAA-Lösungen wurde in einer Rohrmeßstrecke im Bereich ausgebildeter Strömung die Widerstandsverminderung durch Aufnahme der Kurven $\lambda = \lambda\,(Re)$ für Re-Zahlen im Bereich $700 \leqq Re \leqq 10^5$ und Konzentrationen im Bereich $5\ \text{ppm} \leqq c \leqq 50$ ppm gemessen.

Zur Untersuchung der mechanischen Degradation wurden die Lösungen einem definierten Scherfeld im Rohr ausgesetzt. Die hochreinen Laborprodukte boten den Vorteil, daß sie bezüglich ihrer Grenzviskosität eindeutig charakterisierbar und der $\bar{M}_\eta$-Wert damit eindeutig bestimmbar war. Das Zerbrechen der Moleküle konnte so durch Aufnahme der mittleren Molmasse $\bar{M}_\eta$ vor und nach den degradierenden Rohrdurchläufen verfolgt werden.

Ergebnisse

Die Wirkung der Additive ist mit einer Streckung der Moleküle im Scherfeld verbunden. Dabei nehmen diese Energie auf. Es ist zu erwarten, daß bei gegebenem Scherfeld lange Moleküle stärker gedehnt werden als kurze. Moleküle hoher Molmasse sollten daher einerseits eine erhöhte Wirksamkeit zeigen, andererseits aber gegen mechanische Degradation anfälliger sein.

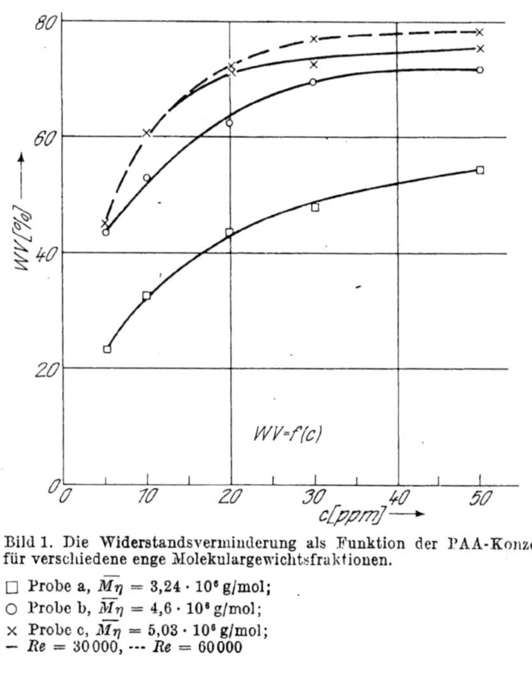

Bild 1. Die Widerstandsverminderung als Funktion der PAA-Konzentration für verschiedene enge Molekulargewichtsfraktionen.

□ Probe a, $\bar{M}_\eta = 3{,}24 \cdot 10^6$ g/mol;
○ Probe b, $\bar{M}_\eta = 4{,}6 \cdot 10^6$ g/mol;
× Probe c, $\bar{M}_\eta = 5{,}03 \cdot 10^6$ g/mol;
— $Re = 30\,000$, --- $Re = 60\,000$

Man erkennt in Bild 1, daß es gelang, hinreichend enge Molmassenfraktionen zu erzeugen, so daß die Molmasse als Parameter deutlich in Erscheinung tritt. Die Widerstandsverminderung WV, definiert durch den Ausdruck

$$WV \equiv \frac{\lambda_w - \lambda}{\lambda_w} \cdot 100\,[\%] \quad \text{mit} \quad \lambda \equiv \frac{2\Delta p}{\varrho u_m^2} \frac{D}{L}$$

ist als Funktion der PAA-Konzentration aufgetragen. Die Re-Zahl ist durch die Beziehung $Re = u_m \cdot D/\nu$ gegeben (L = Rohrlänge entlang derer der Druckabfall Δp gemessen wurde, D = Rohrdurchmesser, u_m = mittlere Strömungsgeschwindigkeit, ν = kinematische Viskosität der Lösung, λ = Widerstandsbeiwert der Lösung).

Die WV-Werte steigen mit der Polymerkonzentration zunächst an und nähern sich oberhalb einer Konzentration, die vom mittleren Molekulargewicht $\bar{M}_\eta$ der Probe abhängt, einem Sättigungswert. Bei gegebener Konzentration ist die Lösung aus der PAA-Probe mit dem höchsten $\bar{M}_\eta$-Wert am wirksamsten. Wie Bild 1 für die PAA-Lösungen der Probe c zeigt, führt eine Erhöhung der Re-Zahl zu einer gesteigerten Widerstandsverminderung, da die Wirksamkeit der Moleküle durch eine stärkere Streckung erhöht wird. Werden die Schubspannungen allerdings zu groß, dann zerbrechen die Moleküle. Überwiegt dieser Vorgang der mechanischen Degradation, so nimmt die Widerstandsverminderung wieder ab.

Die Proben a und c wurden bei $\Delta p/L = 0{,}45$ bar/m, die Probe b bei $\Delta p/L = 0{,}52$ bar/m degradiert. Im letzteren Fall war der Rohreinlauf so „glatt“, daß der laminar-turbulente Umschlag bei $Re \cong 3500$ eintrat, während er sonst bei $Re = 2300$ stattfand.

Das Bild 2 zeigt, daß die langen Moleküle rascher zerbrechen als die kürzeren und die mechanische Degradation während der ersten Rohrdurchläufe besonders stark ist, der $\bar{M}_\eta$-Wert sich anschließend aber nur noch langsam ändert.

Wie Bild 3 zeigt, rufen die in gleicher Weise degradierten Lösungen je nach Re-Zahl der Rohrströmung unterschiedliche Änderungen der Widerstandsverminderung hervor.

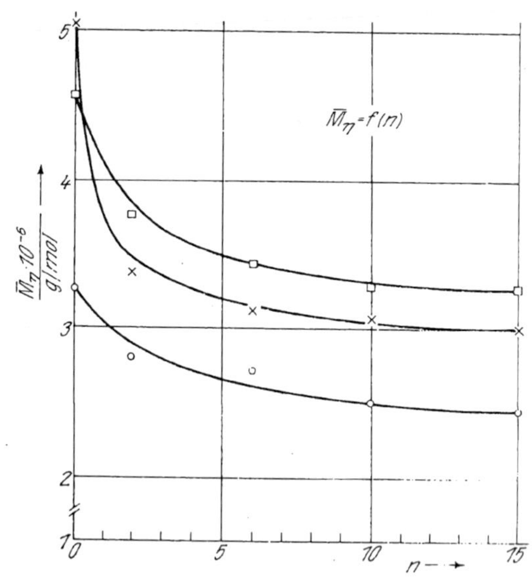

Bild 2. Die Änderung der mittleren Molmasse $\bar{M}_\eta$ während der Degradation (○ Probe a, □ Probe b, × Probe c), c = 50 ppm.

Bei höherer Re-Zahl, d. h. in starken Scherfeldern werden auch kürzere noch nicht degradierte Moleküle hinreichend gestreckt, um widerstandsvermindernd wirksam zu sein. Die Auswirkung einer gegebenen Degradation ist daher für kleinere Re-Zahlen größer. Die WV-Werte der PAA-Probe mit der höheren Molmasse nehmen insbesondere während der ersten Rohrdurchläufe stärker ab als die der Probe mit der niedrigeren Molmasse, siehe auch Bild 2. Bei einer Re-Zahl von 60000 unterscheiden sich die beiden PAA-Proben in ihrer Wirkung zunächst um etwa 20%, aber schon nach wenigen Rohrdurchläufen nur noch um etwa 5%.

Die mittlere Molmasse der frischen, niedermolekularen Probe a wird von der höhermolekularen Probe c nach einigen Rohrdurchläufen erreicht bzw. unterschritten, siehe Bild 2. Bei

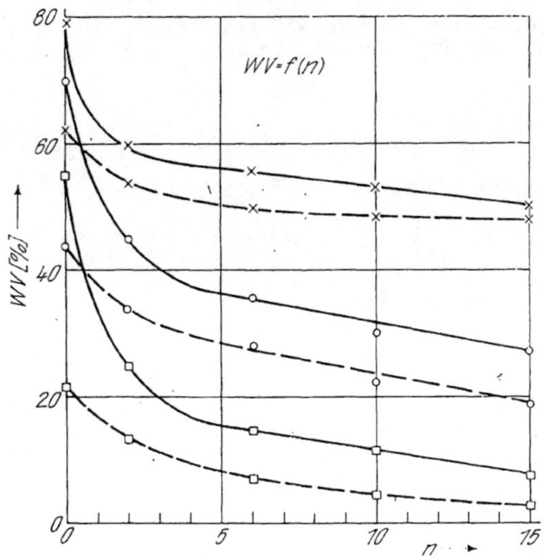

Bild 3. Auswirkung gegebener Degradationen auf die WV-Werte in Abhängigkeit von der Re-Zahl der Rohrströmung (× $Re = 60000$, ○ $Re = 20000$, □ $Re = 10000$; — Probe c, --- Probe a), $c = 50$ ppm.

gleichem $\overline{M}_\eta$-Wert ergibt sich für die frische Lösung eine höhere Widerstandsverminderung als für die degradierte Lösung. Es erscheint daher von Wichtigkeit, die Änderung der Molmassenverteilung mit der Degradation zu verfolgen.

Die Verfasser danken den Herren Prof. Dr. J. KLEIN und Priv.-Doz. Dr. W.-M. KULICKE vom Institut für Chemische Technologie der TU Braunschweig für die großzügige Hilfe bei der Herstellung der PAA-Proben.

Literatur

1 BERMAN, N. S., Reduction of the highest molecular weight fraction of polyethyleneoxid, Phys. Fluid, Vol. 20, No. 5, 1977, pp. 715–719.

Eingereicht am 11. Juni 1981 (Vortrag GAMM-Tagung Würzburg 1981)

Anschrift: Prof. Dr. B. GAMPERT, Dipl.-Ing. P. WAGNER, Fachbereich 12, Universität Essen-Gesamthochschule, D 4300 Essen 1, Schützenbahn 70, BRD

ZAMM 62, 269–272 (1982)

C. Y. WANG

Equilibrium Shapes and Stability of a Rotating Flexible Shaft with an End Mass

This paper studies the axial rotation of an elastic shaft with an end mass. The problem depends heavily on a non-dimensional parameter J which measures the relative importance of the centrifugal force to the elastic restoring force. The shaft remains straight for $J < 3$. Two equilibrium solutions exist for $J > 3$ with the bent shaft as the stable one. The equilibrium shapes for large deflections are found by numerical integration. Asymptotic solution for large J is also presented.

Formulation

Consider an originally straight rotating shaft, one end in a bushing and the other end is attached to a concentrated mass m (Fig. 1a). If the rotation angular velocity Ω is large enough the centrifugal force F will bend the shaft into a new equilibrium shape (Fig. 1b). The present paper studies the stability and equilibrium shapes of such a shaft.

We assume the following: 1) The mass of the shaft is negligible in comparison with the end mass. 2) The gravitational force is small in comparison with the centrifugal force or the elastic restoring force of the shaft. 3) The diameter to length ratio is small enough such that the shaft may be considered to be an elastica.

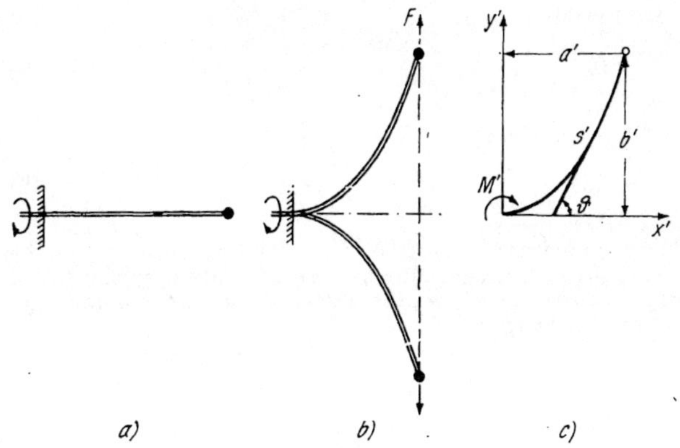

Fig. 1. The coordinate system

Using the coordinate system shown in Fig. 1c the elastica equations are

$$EI \frac{d\theta}{ds'} = M' - Fx' , \tag{1}$$

$$\frac{dx'}{ds'} = \cos \theta , \qquad \frac{dy'}{ds'} = \sin \theta \tag{2}$$

where θ is the local angle of inclination, s' is the arc length from the fixed end, x', y' are cartesian coordinates, EI is the flexural rigidity, and M' is the maximum moment. We shall normalize all lengths by the shaft length L, the moment by EI/L, the force by EI/L^2 and drop primes. Eqs. (1, 2) become

$$\frac{d\theta}{ds} = M - Jbx , \qquad \frac{dx}{ds} = \cos \theta , \qquad \frac{dy}{ds} = \sin \theta , \tag{3), (4}$$

where $b = y(1)$ and $J \equiv m\Omega^2 L^3/EI$ is a nondimensional number which measures the importance of centrifugal effect to elastic rigidity. Differentiating once, the problem becomes

$$\frac{d^2\theta}{ds^2} = -Jb \cos \theta , \qquad \frac{dx}{ds} = \cos \theta , \qquad \frac{dy}{ds} = \sin \theta , \tag{5), (6}$$

with the boundary conditions

$$\theta(0) = 0 , \qquad \frac{d\theta}{ds}(1) = 0 , \qquad x(0) = y(0) = 0 , \qquad y(1) = b. \tag{7}$$

The solution to Eq. (5) may be expressed in terms of elliptic integrals

$$s(\theta) = \frac{1}{\sqrt{Jb}} \left\{ F\left[\sqrt{\frac{1 + \sin \beta}{2}} , \sin^{-1} \left(\sqrt{\frac{2}{1 + \beta}} \sin \left(\frac{\theta}{2} + \frac{\pi}{4}\right)\right)\right] - F\left[\sqrt{\frac{1 + \sin \beta}{2}} , \sin^{-1} \left(\frac{1}{\sqrt{1 + \sin \beta}}\right)\right]\right\} \tag{8}$$

where the unknowns β and b are to be obtained from the relationships

$$\beta = \theta(1) , \qquad b = \int_0^1 \sin \theta \, ds . \tag{9}$$

However, it is almost impossible to invert Eq. (8) for the necessary evaluation. We feel the character of the solution may be more readily revealed by not using elliptic functions.

The Only Eigenvalue

We notice that the straight shaft

$$\theta = b = y = 0 , \qquad x = s \tag{10}$$

is always a solution. Let us investigate the possibility of non-trivial solutions for small θ. We set

$$\sin \theta \approx \theta , \qquad \cos \theta \approx 1 . \tag{11}$$

Eq. (5) can be integrated twice to give

$$\theta = -Jb \left(\frac{s^2}{2} - s\right) . \tag{12}$$

Eq. (6) yields

$$y = -Jb\left(\frac{s^3}{6} - \frac{s^2}{2}\right). \tag{13}$$

At $s = 1$ Eq. (13) gives

$$b = \frac{Jb}{3}. \tag{14}$$

This shows either $b = 0$ (the trivial solution) or $J = 3$ as the sole eigenvalue where solutions may bifurcate from the trivial one. As we shall see later, the trivial solution is the unique and stable solution for $J < 3$.

Asymptotic Solutions for Large J

Large J means large end mass, high rotation rate, long shaft or small flexural rigidity. We set

$$J \equiv \frac{1}{\varepsilon^2} \gg 1 \tag{15}$$

where ε is a small number. Eq. (5) then becomes a singular perturbation problem

$$\varepsilon^2 \frac{d^2\theta}{ds^2} = -b\cos\theta. \tag{16}$$

From physical considerations, we limit $0 \leqq \theta \leqq \frac{\pi}{2}$. The interior solution is

$$\theta = \frac{\pi}{2}. \tag{17}$$

In the region near $s = 0$ we set

$$s = \varepsilon t, \tag{18}$$

$$\theta = \theta_0(t) + \varepsilon\theta_1(t) + \dots, \qquad b = b_0 + \varepsilon b_1 + \dots \tag{19), (20}$$

and obtain the successive equations

$$\frac{d^2\theta_0}{dt^2} = -b_0\cos\theta_0, \tag{21}$$

$$\frac{d^2\theta_1}{dt^2} = b_0\theta_1\sin\theta_0 - b_1\cos\theta_0 \tag{22}$$

with the boundary conditions

$$\theta_0(0) = 0, \quad \theta_0(\infty) \to \frac{\pi}{2}, \qquad \theta_1(0) = 0, \quad \theta_1(\infty) \to 0. \tag{23), (24}$$

We multiply Eq. (21) by $d\theta_0/dt$ and integrate to obtain

$$\frac{1}{2}\left(\frac{d\theta_0}{dt}\right)^2 = -b_0\sin\theta_0 + b_0 \tag{25}$$

where the constant of integration is obtained from the boundary condition at infinity. Thus

$$\frac{d\theta_0}{dt} = \sqrt{2b_0(1-\sin\theta_0)} = 2\sqrt{b_0}\sin\left(\frac{\pi - 2\theta_0}{4}\right). \tag{26}$$

One further integration gives

$$t = -b_0^{-1/2}\ln\frac{\tan\left(\frac{\pi - 2\theta_0}{8}\right)}{\tan\left(\frac{\pi}{8}\right)}. \tag{27}$$

To determine b_0 we write

$$\frac{dy}{ds} = \sin\theta = 1 - 2\sin^2\left(\frac{\pi - 2\theta}{4}\right). \tag{28}$$

Integrate once

$$y = s + O(\varepsilon). \tag{29}$$

The boundary condition at $s = 1$ gives

$$b_0 = 1. \tag{30}$$

It is very difficult to solve for the next order correction θ_1 from Eqs. (22) and (27). However, Eq. (27) suggests the transformation

$$\theta = 4\tan^{-1}u + \frac{\pi}{2}. \tag{31}$$

Using Eq. (31) and starting afresh, Eq. (16) becomes

$$(1 + u^2)\frac{d^2u}{dt^2} - 2u\left(\frac{du}{dt}\right)^2 - bu(1 - u^2) = 0 \tag{32}$$

with the boundary conditions

$$u(0) = -\tan\frac{\pi}{8} \equiv c, \qquad u(\infty) = 0. \tag{33}$$

We then expand

$$u = u_0(t) + \varepsilon u_1(t) + \dots \tag{34}$$

Eqs. (20, 30, 32—34) yield

$$(1 + u_0^2)\frac{d^2u_0}{dt^2} - 2u_0\left(\frac{du_0}{dt}\right)^2 - u_0(1 - u_0^2) = 0, \tag{35}$$

$$u_0(0) = c, \qquad u_0(\infty) = 0, \tag{36}$$

$$(1 + u_0^2)\frac{d^2u_1}{dt^2} - 4\left(u_0\frac{du_0}{dt}\right)\frac{du_1}{dt} + \left[2u_0\frac{d^2u_0}{dt^2} - 2\left(\frac{du_0}{dt}\right)^2 + 3u_0^2 - 1\right]u_1 = b_1u_0(1 - u_0^2),$$

$$u_1(0) = 0, \qquad u_1(\infty) = 0. \tag{37}$$

Guided by Eq. (27), the zeroeth order solution is

$$u_0 = c\,e^{-t}. \tag{38}$$

Eq. (37) reduces to

$$(1 + c^2e^{-2t})\frac{d^2u_1}{dt^2} + 4c^2e^{-2t}\frac{du_1}{dt} + (3c^2e^{-2t} - 1)\,u_1 = b_1c\,e^{-t}(1 - c^2e^{-2t}). \tag{39}$$

The solution is

$$u_1 = -\frac{b_1c}{2}\,t\,e^{-t}. \tag{40}$$

To determine b_1, we set

$$y = s + \varepsilon y_1(t) + \dots \tag{41}$$

Using Eqs. (28, 31, 41) we find

$$\frac{dy_1}{dt} = -2\sin^2\left(\frac{\pi - 2\theta_0}{4}\right) = \frac{-8u_0^2}{(1 + u_0^2)^2}. \tag{42}$$

The solution is

$$y_1 = \frac{-4}{1 + c^2e^{-2t}} + \frac{4}{1 + c^2}. \tag{43}$$

Therefore

$$b_1 = y_1(\infty) = \frac{-4c^2}{1 + c^2} = \sqrt{2} - 2. \tag{44}$$

The maximum moment is

$$M = \left.\frac{d\theta}{ds}\right|_0 = \left.\frac{1}{\varepsilon}\frac{d\theta}{du}\frac{du}{dt}\right|_0 = \frac{\sqrt{2}}{\varepsilon}\left[1 - \varepsilon\left(1 - \frac{1}{\sqrt{2}}\right) + O(\varepsilon^2)\right]. \tag{45}$$

From Eq. (3) the axial distance is

$$a = \frac{M}{Jb} = \varepsilon\sqrt{2}\left[1 - 3\varepsilon\left(1 - \frac{1}{\sqrt{2}}\right) + O(\varepsilon^2)\right]. \tag{46}$$

Numerical Integration

Exact solutions may be obtained by numerically integrating Eqs. (5—7). This difficult two-point boundary value problem may be further simplified by the following transform. Let

$$r = \sqrt{Jb}\,(1 - s)\,, \tag{47}$$

$$\eta = \sqrt{Jb}\,(b - y)\,. \tag{48}$$

Then the governing equations become

$$\frac{d^2\theta}{dr^2} = -\cos\theta\,, \qquad \frac{d\eta}{dr} = \sin\theta\,. \tag{49), (50}$$

Starting from $r = 0$, we set $d\theta/dr = \eta = 0$ and choose any starting value for θ. Then we integrate Eqs. (49, 50) as an initial value problem by the fourth order RUNGE-KUTTA algorithm. The integration terminates at $\theta = 0$ where $r = r^*$, say. Then

$$J = (r^*)^3/\eta(r^*)\,, \qquad b = \eta(r^*)/r^*\,, \tag{51), (52}$$

$$a = -\frac{d\theta}{dr}(r^*)/r^*\,, \qquad M = -r^*\frac{d\theta}{dr}(r^*)\,. \tag{53), (54}$$

In order to obtain the equilibrium shape we integrate Eqs. (5–7) with the newly found initial value

$$\frac{d\theta}{ds}(0) = -\sqrt{Jb}\,\frac{d\theta}{dr}(r^*)\,, \tag{55}$$

where b is given by Eq. (52).

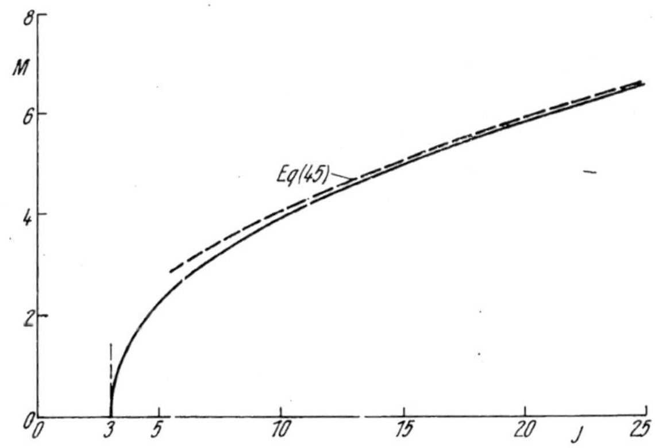

Fig. 2. The maximum moment

Fig. 2 shows the moment M obtained by this numerical scheme. It is seen that the trivial solution is the unique solution for $J < 3$ and two equilibrium solutions exist for $J > 3$. Also shown are the approximate analytic solutions. The two term expansion for the maximum moment M, Eq. (45), is within 5% of the exact value if $J > 9$. Fig. 3 shows the location (a, b) of the tip mass. The approximation for a is not as accurate as that for b.

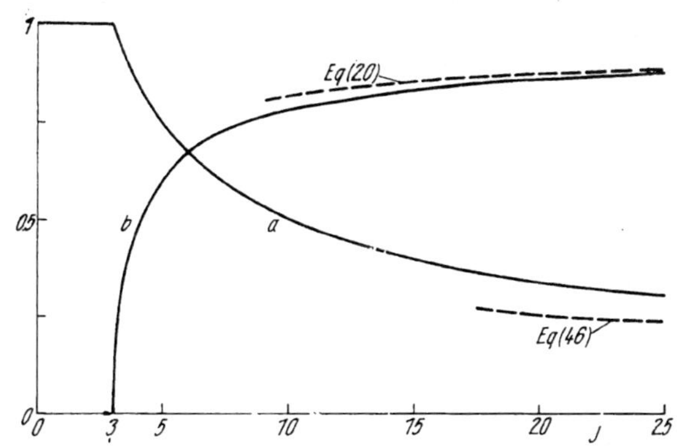

Fig. 3. The distances a and b

Fig. 4 shows some nontrivial equilibrium shapes for various J. We see that the shape is very sensitive to J when it is closed to, but larger than 3. The boundary layer character is also seen for large J.

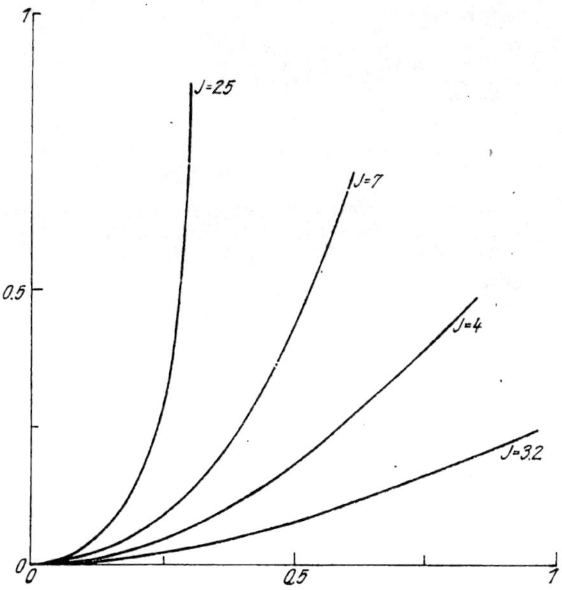

Fig. 4. Some equilibrium shapes

Discussion

The present problem is related to the large deflection of a horizontal cantilever studied by BISSHOPP and DRUCKER [1]. There are, however, significant differences. The coupling of y and θ through the unknown b in our problem makes solutions by elliptic functions impractical. Also, the present problem is an eigenvalue problem with non unique solutions. The cantilever problem is not. It is interesting that in our case a single eigenvalue is found while the usual elastica eigenvalue problems, such as buckling [2], involve infinite number of eigenvalues.

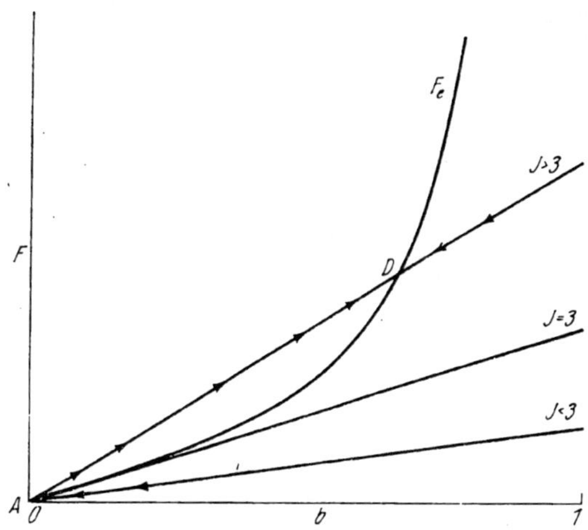

Fig. 5. Elastic restoring force F_e and centrifugal forces as a function of displacement b

Since non uniquenesss occur for $J > 3$, we shall determine which equilibrium solution is stable. Fig. 5 shows force plotted as a function of distance b. The curve F_e represents the elastic restoring force of a cantilever [1]. Centrifugal forces are straight lines from the origin. Equilibrium occurs when the centrifugal force balances the elastic force. There is only one solution (the trivial solution at A) for $J < 3$. Two solutions (at A and D) occur when $J > 3$. The solution at A is unstable since, for given $J > 3$, any small departure of b from zero will lead to a non equilibrium state with larger centrifugal force than elastic restoring force. This promotes a larger b which eventually ends at state D. On the other hand, the forces work opposite for displacements larger than that of state D. Thus D is the only stable equilibrium state for $J > 3$.

References

1 Bisshopp, K. E.; Drucker, D. C., Large Deflection of Cantilever Beams, Quart. Appl. Math., 3, 272–275 (1945).
2 Timoshenko, S. P.; Gere, J. M., Theory of Elastic Stability, McGraw-Hill, New York, (1961).

Eingereicht am 17. 12. 1980

Anschrift: C. Y. Wang, Professor of Applied Mathematics, Michigan State University, East Lansig, Michigan 48824, USA

ZAMM 62, 272–274 (1982)

S. M. Sinha / D. R. Aylawadi

Optimality Criteria for a Class of Non-differentiable Mathematical Programs

Introduction

Kuhn and Tucker [4] obtained a set of necessary optimality conditions under constraint qualification for differentiable non-linear programs. They also showed that if the functions defining the program are convex, the conditions are also sufficient for minimization problems. But in the absence of convexity one is not sure of the sufficient criteria. Moreover, when one or more of the functions defining the program are not differentiable, there is no general method to find the optimality criteria, whether necessary or sufficient. However, saddle point optimality conditions are available for non-differentiable functions. Optimality conditions for programs with non-differentiable objective functions have been studied by Misra [5], Mond [6]. Sinha [11] discussed the solution of a special class of such problems. Bazaraa [1] obtained optimality criteria without differentiability using the concept of polar cones.

Here we deal with a mathematical program where one of the constraining functions is non-differentiable and find necessary and sufficient optimality conditions. The analysis is then extended to programs with more than one non-differentiable constraint function.

The problem under consideration is

(P) Maximize $f(X)$ (1)

subject to $g_i(X) \leqq 0; \quad i = 1, 2, \dots, m,$ (2)

$$p^{\mathsf T}X + (X^{\mathsf T}BX)^{1/2} \leqq 1\,, \tag{3}$$

where f is a concave and differentiable function, g_i, $i = 1, 2, \dots, m$, are convex and differentiable functions, B is an $n \times n$ symmetric positive semi-definite matrix and $X = (x_1, x_2, \dots, x_n)^{\mathsf T}$ is an n-dimensional vector ("T" will indicate the transpose of a vector or a matrix).

A stochastic programming problem in which the coefficients of one of the constraining functions are random variables, of which only the means, variances and co-variances are known can be reduced to this form [11].

Let the constraint set be denoted by C_P, i.e.

$$C_P = \{X \mid g_i(X) \leqq 0, i = 1, 2, \dots, m; p^{\mathsf T}X + (X^{\mathsf T}BX)^{1/2} \leqq 1\}\,.$$

We define

$$g(X) = [g_1(X), g_2(X), \dots, g_m(X)]^{\mathsf T}\,,$$

and the operator

$$\nabla = \left(\frac{\partial}{\partial x_1}, \frac{\partial}{\partial x_2}, \dots, \frac{\partial}{\partial x_n}\right)^{\mathsf T}.$$

We assume that the set C_P is non-empty.

Some useful results

In this section we shall state some known results as lemmas which will be needed for our analysis.

Lemma 1: *The set C_P is convex* [11].

Hence problem (P) requires maximization of a concave function over a convex set and so any local maximum is a global maximum.

Lemma 2 (Eisenberg [3]): *If $X, Y \in E^n$ and B is an $n \times n$ symmetric positive semi-definite matrix, then*

$$X^{\mathsf T}BY \leqq (X^{\mathsf T}BX)^{1/2}\,(Y^{\mathsf T}BY)^{1/2}$$

and equality holds, if $BX = \lambda BY$; $\lambda \geqq 0$.

It follows easily from lemma 2 that $X^{\mathsf T}BX = 0$ implies $BX = 0$.

Lemma 3 (Farkas Lemma): *Let A be an $m \times n$ matrix and C be an n-component vector, then $AX \geqq 0$ implies $C^{\mathsf T}X \geqq 0$ for some $X \in E^n$ if and only if $A^{\mathsf T}Y = C$ for some $Y \in E^m_+$.*

Lemma 4 (Motzkin [8]): *Let M_1, M_2, M_3 be real constant matrices with M_1 being non-empty, then either the system*

$$Y^{\mathsf T}M_1 > 0; \quad Y^{\mathsf T}M_2 \geqq 0; \quad Y^{\mathsf T}M_3 = 0; \quad Y \in E^n\,,$$

has a solution of the system

$$M_1Z_1 + M_2Z_2 + M_3Z_3 = 0\,,$$

$$Z_1 > 0; \quad Z_2 \geqq 0$$

has a solution, but not both.

Notations and preliminaries

Let X_0 be an optimal solution of the problem (P). We define the set $Q_0 = \{i \mid g_i(X_0) = 0\}$ and the set

$$Z_0 = \Big\{Z \mid (\nabla g_i(X_0))^{\mathsf T} Z \leqq 0; \quad i \in Q_0\,, \tag{4}$$

$$p^{\mathsf T}Z + \frac{Z^{\mathsf T}BX_0}{(X_0^{\mathsf T}BX_0)^{1/2}} \leqq 0 \quad \text{if} \quad X_0^{\mathsf T}BX_0 > 0 \quad \text{and}$$

$$p^{\mathsf T}X_0 + (X_0^{\mathsf T}BX_0)^{1/2} = 1\,, \tag{5}$$

$$p^{\mathsf T}Z + (Z^{\mathsf T}BZ)^{1/2} \leqq 0 \quad \text{if} \quad X_0^{\mathsf T}BX_0 = 0 \quad \text{and}$$

$$p^{\mathsf T}X_0 = (X_0^{\mathsf T}BX_0)^{1/2} = 1; \tag{6}$$

$$(\nabla f(X_0))^{\mathsf T} Z > 0\Big\}\,. \tag{7}$$

The assumption that the set Z_0 is empty consists of two parts:

a) If $X_0^{\mathsf T}BX_0 > 0$ and $p^{\mathsf T}X_0 + (X_0^{\mathsf T}BX_0)^{1/2} = 1$ then for every Z satisfying (4) and (5), it is true that

$$(\nabla f(X_0))^{\mathsf T} Z \leqq 0\,. \tag{8}$$

b) If $X_0^{\mathsf T}BX_0 = 0$ and $p^{\mathsf T}X_0 + (X_0^{\mathsf T}BX_0)^{1/2} = 1$ then for every vector Z satisfying (4) and (6) it is true that

$$(\nabla f(X_0))^{\mathsf T} Z \leqq 0\,.$$

Note that a) is the usual constraint qualification when the constraint $p^{\mathsf T}X + (X^{\mathsf T}BX)^{1/2} \leqq 1$ is differentiable at X_0 (in this case we have $X_0^{\mathsf T}BX_0 > 0$), while 6) is the requirement that the directional derivative (sub-gradient) of $p^{\mathsf T}X + (X^{\mathsf T}BX)^{1/2}$ in the direction of the vector Z is negative.

Necessary optimality conditions

Theorem 1: *If X_0 solves the problem (P) and the corresponding set Z_0 is empty, then there exist $Y_0 \in E^m_+$, $W_0 \in E^n$ and a scalar $\alpha_1^0 \geqq 0$ such that $(X_0, Y_0, \alpha_1^0, W_0)$ satisfies*

$$Y^{\mathsf T}g(X)\,, \tag{9}$$

$$\alpha_1(1 - p^{\mathsf T}X - (X^{\mathsf T}BX)^{1/2}) = 0\,, \tag{10}$$

$$Y^{\mathsf T}\nabla g(X) + \alpha_1 p + BW = \nabla f(X)\,, \tag{11}$$

$$W^{\mathsf T}BX = \alpha_1(X^{\mathsf T}BX)^{1/2}\,, \tag{12}$$

$$Y, \alpha_1 \geqq 0\,. \tag{13, 14}$$

Proof:

Case I: $X_0^{\mathsf T}BX_0 > 0$. If $p^{\mathsf T}X_0 + (X_0^{\mathsf T}BX_0)^{1/2} = 1$ and the set Z_0 is empty then a Z satisfying (4) and (5) implies relation (8). By Lemma 3 it then follows that there exist $\bar y_i \geqq 0$; $i \in Q_0$, and

$\hat{\alpha}_1 \geqq 0$ satisfying

$$-\sum_{i \in Q_0} \bar{y}_i \nabla g_i(X_0) - \bar{\alpha}_1 \left(p + \frac{BX_0}{(X_0^\top BX_0)^{1/2}}\right) = -\nabla f(X_0) . \tag{15}$$

If $p^\top X_0 + (X_0^\top BX_0)^{1/2} < 1$, the assumption of emptiness of the set Z_0 then gives that for Z satisfying (4) the relation (8) is true. By Lemma 3 there exist $\bar{y}_i^* \geqq 0$; $i \in Q_0$, satisfying

$$-\sum_{i \in Q_0} \bar{y}_i^* \nabla g_i(X_0) = -\nabla f(X_0) . \tag{16}$$

Define

$$y_i = \begin{cases} \bar{y}_i & i \in Q_0 \text{ if } p^\top X_0 + (X_0{}^\top BX_0)^{1/2} = 1 , \\ \bar{y}_i^* & i \in Q_0 \text{ if } p^\top X_0 + (X_0{}^\top BX_0)^{1/2} < 1 , \\ 0 & \text{otherwise} , \end{cases} \tag{17}$$

$Y = (y_1, y_2, \dots, y_m)^\top$,

$$\alpha_1 = \begin{cases} \bar{\alpha}_1 & \text{if } p^\top X_0 + (X_0^\top BX_0)^{1/2} = 1 , \\ 0 & \text{otherwise} \end{cases} \tag{18}$$

and

$$W = \frac{\alpha_1 X_0}{(X_0^\top BX_0)^{1/2}} . \tag{19}$$

Thus there exist $Y \in E_+^m$; $W \in E^n$ and scalars $\alpha_1 \geqq 0$ such that (X_0, Y, α_1, W) satisfy relations (9)—(14).

Case II: $X_0^\top BX_0 = 0$ (the function $(X^\top BX)^{1/2}$ is therefore not differentiable at X_0).

If $p^\top X_0 + (X_0^\top BX_0)^{1/2} = 1$ and the set Z_0 is empty then Z satisfies (4) and (6) implies (8).

We now require to prove the following lemma.

Lemma 5: *Let A be an $m \times n$ matrix and C be an $n \times 1$ vector; B is an $n \times n$ symmetric positive semi-definite matrix. If*

$$\left.\begin{array}{l} AX \leqq 0; \quad p^\top X + (X^\top BX)^{1/2} \leqq 0; \quad X \in E^n \text{ implies} \\ \qquad\qquad\qquad C^\top X \leqq 0 \end{array}\right\} \text{System-I}$$

then there exists $Y \in E_+^m$; $W \in E^n$ and a scalar α_1 satisfying

$$\left.\begin{array}{ll} A^\top Y + \alpha_1 p + BW = C & \text{(i)} , \\ Y, \alpha_1 \geqq 0 & \text{(ii)} . \end{array}\right\} \text{System-II}$$

Proof of the Lemma: System I must imply System II. If not, i.e. there is no solution to

$$A^\top Y + \alpha_1 p + BW = C; \quad Y \geqq 0; \quad \alpha_1 \geqq 0 ,$$

i.e.

$$A^\top Y + \alpha_1 p + BW - \gamma C = 0; \quad Y \geqq 0; \quad \alpha_1 \geqq 0; \quad \gamma > 0 .$$

By Motzkin's theorem (Lemma 4), there exists a Z such that

$$AZ \geqq 0; \quad p^\top Z \geqq 0; \quad BZ = 0; \quad -C^\top Z > 0 .$$

Setting $X = -Z$ yields

$$AX \leqq 0; \quad p^\top X \leqq 0; \quad BX = 0; \quad C^\top X > 0 .$$

From $BX = 0$ follows $(X^\top BX)^{1/2} = 0$; $p^\top X + (X^\top BX)^{1/2} \leqq 0$, what contradicts the non-existence of a solution to System I. Hence the System II has a solution. Q.E.D.

Applying Lemma 5 to the system (4), (6) and (8) we note that there exist $\hat{y}_i \geqq 0$; $i \in Q_0$; $\hat{W} \in E^n$ and a scalar $\hat{\alpha}_1 \geqq 0$ satisfying

$$\sum_{i \in Q_0} \hat{y}_i \nabla g_i(X_0) + \hat{\alpha}_1 p + B\hat{W} = \nabla f(X_0) .$$

Now if $p^\top X_0 + (X_0^\top BX_0)^{1/2} < 1$, the assumption of emptiness of the set Z_0 then gives that Z satisfying (4) implies (8). By Farkas' lemma it then follows the existence of $\hat{y}_i^* \geqq 0$; $i \in Q_0$, satisfying

$$\sum_{i \in Q_0} \hat{y}_i^* \nabla g_i(X_0) = \nabla f(X_0) .$$

Define

$$y_i = \begin{cases} \hat{y}_i, & i \in Q_0, \text{ if } p^\top X_0 + (X_0^\top BX_0)^{1/2} = 1 , \\ \hat{y}_i^*, & i \in Q_0 , \text{ if } p^\top X_0 + (X_0^\top BX_0)^{1/2} < 1 , \\ 0, & \text{otherwise} , \end{cases}$$

$Y = (y_1, y_2, \dots, y_m)^\top$,

$$\alpha_1 = \begin{cases} \hat{\alpha}_1 & \text{if } p^\top X_0 + (X_0^\top BX_0)^{1/2} = 1 , \\ 0 & \text{otherwise} , \end{cases}$$

$$W = \begin{cases} \hat{W} & \text{if } p^\top X_0 + (X_0^\top BX_0)^{1/2} = 1 , \\ \text{any } W \text{ with } BW = 0; & \text{otherwise;} \end{cases} \tag{20}$$

relations (9)—(14) are thus satisfied.

Sufficient optimality conditions

Theorem 2: *If $(X_0, Y_0, \alpha_1^0, W_0)$ satisfies relations (2), (3) and (9)—(14) and if further $W_0^\top BW_0 \leqq \alpha_1^{02}$ then X_0 solves the problem (P).*

Proof: Let X be any feasible solution of (P). Consider

$$\begin{aligned} f(X) - f(X_0) &\leqq (X - X_0)^\top \nabla f(X_0) = \\ &\qquad (f \text{ is concave and differentiable}) \\ &= (X - X_0)^\top (Y_0^\top \nabla g(X_0) + \alpha_1^0 p + BW_0) \leqq \\ &\leqq Y_0^\top (g(X) - g(X_0)) + \alpha_1^0 p^\top X + W_0^\top BX - \\ &\quad - \alpha_1^0 p^\top X_0 - W_0^\top BX_0 = \\ &= Y_0^\top g(X) + \alpha_1^0 p^\top X + W_0^\top BX - \\ &\quad - \alpha_1^0 (p_0^\top X + (X_0^\top BX_0)^{1/2}) \\ &\quad \text{(using (9) and (12))} \\ &\leqq Y_0^\top g(X) + \alpha_1^0 p^\top X + \\ &\quad + (W_0^\top BW_0)^{1/2} (X^\top BX)^{1/2} - \alpha_1^0 \leqq \\ &\quad \text{(using (10) and Lemma 2)} \\ &\leqq Y_0^\top g(X) + \alpha_1^0 p^\top X + \alpha_1^0 (X^\top BX)^{1/2} - \alpha_1^0 \leqq \\ &\quad \text{(by the condition of the theorem)} \\ &\leqq Y_0^\top g(X) + \alpha_1^0 - \alpha_1^0 \leqq \ (X \text{ is feasible for (P)}) \\ &\leqq 0 \qquad\qquad (g(X) \leqq 0 \text{ and } Y_0 \geqq 0). \end{aligned}$$

Thus

$$f(X_0) \geqq f(X) \quad \text{for all} \quad X \in C_P .$$

Generalization

The above analysis is now extended to programs with more than one non-differentiable functions in the constraints.

Consider the problem

$$\text{(PI)} \quad \begin{array}{l} \text{Maximize } f(X) \\ \text{subject to} \\ g_i(X) \leqq 0 , \qquad i = 1, 2, \dots, m_1 , \\ p_j^\top X + (X^\top B_j X)^{1/2} \leqq 1 , \qquad j = 1, 2, \dots, m_2 , \end{array}$$

where B_j, $j = 1, 2, \dots, m_2$, are $n \times n$ symmetric positive semi-definite matrices. The functions f and g_i, $i = 1, 2, \dots, m_1$, are defined in problem (P).

Let X_0 be an optimal solution of the Problem (PI). Define the set Z_0 as follows:

$$Z_0 = \left\{ Z \mid (\nabla g_i(X_0))^\top Z \leqq 0 , \quad i \in Q_0 , \right.$$
$$p_j^\top Z + \frac{Z^\top B_j X_0}{(X_0^\top B_j X_0)^{1/2}} \leqq 0 \text{ if } X_0^\top B_j X_0 > 0$$
$$\text{and } p_j^\top X + (X_0^\top B_j X_0)^{1/2} = 1, \quad j = 1, 2, \dots, m_2 ,$$
$$p_j^\top Z + (Z^\top B_j Z)^{1/2} \leqq 0 \text{ if } X_0^\top B_j X_0 = 0$$
$$\text{and } p_j^\top X_0 + (X_0^\top B_j X_0)^{1/2} = 1, \quad j = 1, 2, \dots, m_2 ,$$
$$\left. \text{and} \quad (\nabla f(X_0))^\top Z > 0 \right\} .$$

Necessary optimality conditions

If X_0 solves the problem (PI) and the corresponding set Z_0 is empty, then there exist $Y_0 \in E_+^m$; $W_j^0 \in E^n$ and scalars $\alpha_{1j}^0 \geqq 0$; $j = 1, 2, \dots, m_2$, such that $(X_0, Y_0, \alpha_{1j}^0, W_j^0; j = 1, 2, \dots, m_2)$

satisfies

$$Y^{\mathsf{T}} g(X) = 0\,, \tag{21}$$

$$\alpha_{1j}(1 - p_j^{\mathsf{T}} X - (X^{\mathsf{T}} B_j X)^{1/2}) = 0\,, \qquad j = 1, 2, \dots, m_2\,, \tag{22}$$

$$Y^{\mathsf{T}} \nabla g(X) + \sum_{j=1}^{m_2} (\alpha_{1j} p_j + B_j W_j) = \nabla f(X)\,, \tag{23}$$

$$W_j^{\mathsf{T}} B_j X = \alpha_{1j} (X^{\mathsf{T}} B_j X)^{1/2}\,, \qquad j = 1, 2, \dots, m_2\,, \tag{24}$$

$$Y \geqq 0; \quad \alpha_{1j} \geqq 0\,, \qquad j = 1, 2, \dots, m_2\,. \tag{25}$$

Sufficient optimality conditions

If $(X_0, Y_0, \alpha_{1j}^0, W_j^0, j = 1, 2, \dots, m_2)$ satisfy the constraints of (PI) and the relations (21)–(25) and further if $W_j^0 B_j W_j^0 \leqq \alpha_{1j}^{02}$; $j = 1, 2, \dots, m_2$, then X_0 solves the problem (PI).

Acknowledgement

We are very thankful to the referee for his valuable suggestions to improve the quality of this communication.

References

1 Bazaraa, M. S.; Goode, J. J.; Shetty, C. M., Optimality criteria in non-linear programming without differentiability, Operations Res. **19** (1), 77–86 (1971).
2 Eisenberg, E., Supports of a convex function, Bull. Amer. Math. Soc. **68**, 192–195 (1962).
3 Eisenberg, E., A gradient inequality for a class of non-differentiable functions, Operations Res. **14**, 157–163 (1966).
4 Kuhn, H. W.; Tucker, A. W., Non-linear Programming, Proc. Second Berkely Symposium on Mathematical Statistics and Probability, University of California Press 1951, pp. 481–492.
5 Misra, A. D., Optimality criteria for a class of non-differentiable programs, Opsearch **15**, 91–106 (1975).
6 Mond, B., A class of non-differentiable mathematical programs, J. Math. Analysis Appl. **46**, 169–174 (1974).
7 Mond, B., Symmetric duality for non-linear programming, Opsearch, **14**, No. 1, 1–10 (1976).
8 Motzkin, T. S., Two consequences of the transposition theorem of linear inequalities, Econometrica **19**, 184–185 (1951).
9 Mangasarian, O. L.; Fromovitz, S., The Fritz John necessary optimality conditions in the presence of equality and inequality constraints, J. Math. Analysis Appl. **17**, 37–47 (1967).
10 Mangasarian, O. L., Duality in non-linear programming, Quart. Appl. Math. **20**, 300–302 (1962).
11 Sinha, S. M., Stochastic Programming, Research Report No. 22, Operations Research Centre, University of California, Berkeley 1963.
12 Sinha, S. M., A duality theorem for non-linear programming, Management Science **12**, 385–390 (1966).
13 Wolfe, P., A duality theorem for non-linear programming, Quart. Appl. Math. **19**, 239–244 (1961).

Eingereicht am 3. 11. 1979, revidierte Fassungen am 21. 1. und 15. 6. 1981

Anschriften: Prof. Dr. S. M. Sinha; Des Raj Aylawadi, Department of Operational Research, Faculty of Mathematics, University of Delhi, Delhi-110007, India

ZAMM 62, 274–276 (1982)

W. Tischer

Zu einigen Fragen der Bestimmung von Impulsantwortfunktionen aus der Wiener-Hopf-Integralgleichung

1. Einleitung

Bei der Identifikation linearer zeitinvarianter Schwingungssysteme tritt das Problem der Ermittlung optimaler Impulsantwortfunktionen auf. Als Ausgangsbeziehung dient dabei in vielen Fällen das Faltungsintegral

$$\Psi(t) = \int_0^\infty K(\tau)\, \Phi(t - \tau)\, d\tau\,, \qquad 0 \leqq t < \infty\,, \tag{1}$$

$K(t)$ bedeutet die Matrix der Impulsantwortfunktionen, $\Phi(t)$ und $\Psi(t)$ sind Funktionen zur Beschreibung der Eingangs- und Ausgangs-Signale. Sie stellt z. B. bei asymptotisch stabilen Systemen die Eingangs-Ausgangs-Beziehung oder in der Theorie der optimalen Filter die Wiener-Hopf-Integralgleichung für zentrierte stationäre Zufallsprozesse dar. Das Ziel der Identifikationsaufgabe besteht in der Bestimmung von $K(t)$ aus (1) bei vollständig bekannten (gemessenen) $\Psi(t)$ und $\Phi(t)$. Über den Zusammenhang zwischen Systemmatrix und $K(t)$ können damit die Dämpfungs- und Steifigkeitsmatrizen des mechanischen Systems berechnet werden. Im folgenden werden *zwei Methoden zur näherungsweisen Lösung der Integralgleichung* (1) untersucht.

1. In [4] wird nach der Zeitdiskretisierung von (1) in einem hinreichend großen Intervall die Impulsantwortfunktion durch Laguerre-Funktionen approximiert, und die unbekannten Entwicklungskoeffizienten werden aus der Minimumbedingung für ein quadratisches Fehlerfunktional bestimmt. Dabei hat die Wahl des Zeitmaßstabes großen Einfluß auf die Qualität der Approximation. In Abschnitt 2. der vorliegenden Arbeit werden die Bedeutung dieses Parameters untersucht und eine Abschätzung für ihn angegeben. Die Ergebnisse wurden mit bekannten Abschätzungen ([1], [3]) verglichen und numerisch an einem Beispiel ausgewertet.

2. Unter bestimmten Bedingungen, die an das Eingangssignal gestellt werden, kann die Impulsantwortfunktion direkt aus der Integralgleichung (1) durch Anwendung eines Differentialoperators berechnet werden. Diese Methode wird in Abschnitt 3 behandelt.

2. Approximation der Impulsantwortfunktion durch Laguerre-Funktionen und Wahl des Time-Scale-Faktors

Wir betrachten ein asymptotisch stabiles System mit n Freiheitsgraden

$$M\ddot{y} + D\dot{y} + Cy = f\,, \qquad y(t_0) = y_0\,, \qquad \dot{y}(t_0) = \dot{y}_0\,, \tag{2}$$

das auf die Form

$$\dot{x} = Ax + b\,, \qquad x(0) = x_0 \tag{3}$$

gebracht wird mit

$$A = \begin{bmatrix} 0 & E \\ -M^{-1}C & -M^{-1}D \end{bmatrix}, \qquad b = \begin{bmatrix} 0 \\ f \end{bmatrix}, \qquad x = \begin{bmatrix} y \\ \dot{y} \end{bmatrix}, \tag{4}$$

Es wird vorausgesetzt, daß das System (2) n verschiedene Eigenwerte besitzt, d. h. für die Eigenwerte von (3) gilt

$$\left.\begin{aligned} &\lambda_k = \delta_j \pm i\omega_j; && k = 1, 2, \dots, 2n\,, \\ &\delta_j < 0 \quad \text{für alle } j; && j = 1, 2, \dots, n\,. \end{aligned}\right\} \tag{5}$$

Für das System (3) gilt die Integralbeziehung (1) mit

$$K(t) = e^{At} = X\, e^{\Lambda t} X^{-1}\,, \tag{6}$$

wobei X eine Modalmatrix ist, deren Spalten die Eigenvektoren darstellen, $\Lambda = \operatorname{diag}\{\lambda_i\}$ und

$$\dot{K}(t) = AK(t)\,, \qquad K(0) = E\,, \tag{7}$$

E ist die Einheitsmatrix. In [4] werden die Gleichung (1) im Intervall $[0, T]$ diskretisiert und K durch im Intervall $[0, T]$ orthonormierte Funktionen $P_{g,N}$ approximiert,

$$K(t) \sim K_r(i) = \sum_{g=0}^{r} B_g P_{g,N}(i)\,, \quad i = 0, 1, \dots, N\,, \quad r \ll N\,, \tag{8}$$

dabei sind g der Polynomgrad, N die Anzahl der Stützstellen und B_g Entwicklungskoeffizienten. Die $P_{g,N}$ wählt man durch Orthonormierung des Funktionssystems

$$\{\varphi_k(i) = i^k\, e^{-\frac{1}{2}\tilde{\alpha}\frac{T}{N}i}\,; \; k = 0, 1, \dots, r;\; \tilde{\alpha} > 0 \text{ und konstant}; \; i = 0, 1, \dots, N\}, \tag{9}$$

im System der äquidistanten Punkte $\{0, 1, \dots, N\}$. Sie stellen das diskrete Analogon zu den Laguerre-Funktionen

$$l_n^{(\alpha)}(t) = \sqrt{2\alpha}\, e^{-\alpha t} \sum_{k=0}^{n} \frac{n!\,(-2)^k}{(n-k)!\,(k!)^2} (\alpha t)^k\,, \qquad n = 0, 1, \dots, \; \alpha < 0, \tag{10}$$

dar (s. [2], [4]). Aus (8), (9) und (10) ist ersichtlich, daß die Approximationsqualität vom Time-Scale-Faktor α abhängt und daß dieser Parameter das Abklingverhalten der Laguerre-Funktionen und damit auch der approximierten Impulsantwortfunktion bestimmt. Deshalb ist es naheliegend, den Time-Scale-Faktor in Beziehung zu den Realteilen der Systemeigenwerte zu bringen. Dazu wird die Gleichung (6) in der Form

$$K(t) = \sum_{k=1}^{2n} e^{\lambda_k t} \tilde{x}_k y_k^{\mathrm{T}}, \tag{11}$$

geschrieben, wobei $\tilde{x}_k$ Spaltenvektoren von X und $\tilde{y}_k$ Zeilenvektoren von X^{-1} sind. Aus (11) erhält man folgende Normabschätzung für $K(t)$:

$$||K(t)|| \leqq c_1 \, e^{-\{\min\limits_j |\delta_j|\}\, t}, \qquad c_1 = \text{sonst.}; \tag{12}$$

hier ist die Matrizennorm definiert durch

$$||K|| = \sqrt{\operatorname{tr} K^{\mathrm{T}} K}\,. \tag{13}$$

Aus (12) und den oben angeführten Überlegungen läßt sich die Feststellung ableiten, daß man den Time-Scale-Faktor mit dem kleinsten Absolutbetrag der Realteile der Systemeigenwerte gleichsetzen kann. Da jedoch bei Identifikationsaufgaben mit der unbekannten Systemmatrix deren Eigenwerte ebenfalls unbekannt sind, wird eine Abschätzung der Realteile der Eigenwerte und damit des Parameters α aus der Beziehung (1) mit speziellen Eingangssignalen der Art

$$\Phi(t) = e^{-a|t|}, \qquad a > 0, \tag{14}$$

und

$$\Phi(t) = e^{-a|t|} \cos \beta t, \qquad a > 0, \qquad \beta > 0, \tag{15}$$

abgeleitet. Nach Einsetzen von (11) und (14) in (1) und Auswerten der Integrale folgt

$$\Psi(t) = \sum_{k=1}^{2n} \left(\frac{-2a}{\lambda_k^2 - a^2} e^{\lambda_k t} - \frac{1}{\lambda_k + a} e^{-at} \right) \tilde{x}_k \tilde{y}_k^{\mathrm{T}}, \tag{16}$$

und eine Normabschätzung von (16) führt auf die Ungleichung

$$||\Psi(t)|| \leqq \sum_{k=1}^{2n} \left(\frac{2a}{|\lambda_k^2 - a^2|} |e^{\lambda_k t}| + \frac{1}{|\lambda_k + a|} e^{-at} \right) ||\tilde{x}_k \tilde{y}_k^{\mathrm{T}}||\,. \tag{17}$$

Unter Ausnutzung der Eigenschaften der Modalmatrix und mit (9) folgt aus (17) die Abschätzung

$$||\Psi(t)|| \leqq c_2 \, e^{-\alpha_E t} + c_3 \, e^{-at} \tag{18}$$

mit $c_1 = \text{const.}$, $c_2 = \text{const.}$ und

$$\alpha_E = \min_{(j)} |\delta_j|\,. \tag{19}$$

Die Abschätzung (18) bildet die gesuchte Beziehung zur näherungsweisen Bestimmung des kleinsten Absolutbetrages der Realteile der Systemeigenwerte bei bekannten Eingangs- und Ausgangsgrößen a und $\Psi(t)$. Es sei bemerkt, daß man mit (15) ebenfalls die Beziehung (18) erhält.

Zur optimalen Wahl von α wird in [3] eine Formel angegeben. Ausgehend von der Reihenentwicklung

$$K(t) = \sum_{n=0}^{\infty} C_n l_n^{(\alpha)}(t) \tag{20}$$

wird der mittlere quadratische Approximationsfehler

$$\varepsilon_N^2(\alpha) = \frac{||K(t) - \sum\limits_{n=0}^{N-1} C_n l_n^{(\alpha)}(t)||^2_{L^2[0,\infty)}}{||K(t)||^2_{L^2[0,\infty)}} \tag{21}$$

minimiert, woraus die folgende Formel für den Time-Scale-Faktor folgt:

$$\alpha_p = \frac{||t^{1/2} \dot{K}(t)||_{L^2[0,\infty)}}{||t^{1/2} K(t)||_{L^2[0,\infty)}}. \tag{22}$$

Da jedoch $K(t)$ erst im Ergebnis der Identifikation erhalten wird, ist die Formel (22) nicht direkt anwendbar, kann aber bei bestimmten Eingangssignalen so modifiziert werden, daß nur bekannte Größen zur Abschätzung von α benötigt werden (siehe dazu Abschnitt 3.).

Der Zusammenhang zwischen α und der Systemmatrix bzw. deren Eigenwerten wird nach Einsetzen von (7) in (22) sichtbar:

$$\alpha_p \leqq ||A|| \tag{23}$$

mit der Matrixnorm (13).

Am Beispiel des Einmassenschwingers werden nun die durchgeführten Betrachtungen verdeutlicht. Für die skalare Differentialgleichung

$$\ddot{y} + 2d\omega \dot{y}_0 + \omega_0^2 y = f, \qquad \omega_0 > 0, \qquad 0 < d < 1, \tag{24}$$

mit exponentiell korrellierter Eingangsgröße f, d. h.

$$R_{ff}(t - \tau) = e^{-a|t-\tau|} = \Phi(t - \tau), \tag{25}$$

erhält man

$$K(t) = e^{-d\omega_0 t} \frac{1}{\omega_0 \sqrt{1 - d^2}} \sin \omega_0 \sqrt{1 - d^2}\, t\,. \tag{26}$$

Nach der in [4] angeführten Methode wurde $K(t)$ identifiziert.

Die Abhängigkeit der beiden Fehlerfunktionale

$$\varepsilon_1^2 = ||\Psi(t) - \int_0^T \bar{K}(\tau)\, \Phi(t - \tau)\, d\tau||^2_{L^2[0,T)}, \tag{27}$$

$$\varepsilon_2^2 = ||K - \bar{K}||^2_{L^2[0,T]}, \tag{28}$$

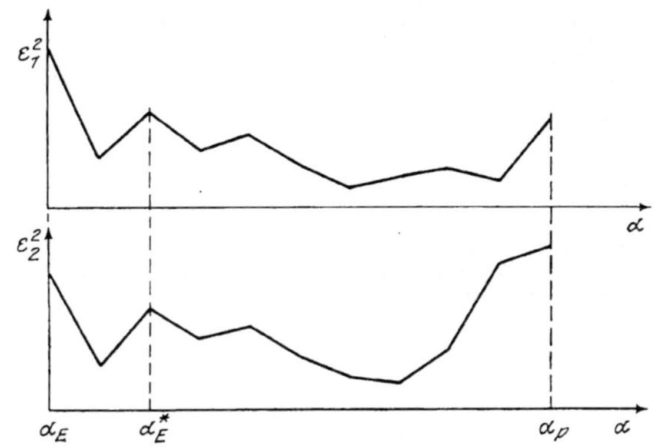

Abb. 1. Abhängigkeit der Approximationsfehler von α

von α ist in Abb. 1 dargestellt ($\bar{K}$ ist approximierte Funktion).

Mit $\alpha_E = d\omega_0$ wurde der Absolutbetrag des Realteiles des Eigenwertes bezeichnet, α_p wurde aus der Formel (22) und α_E^* aus (18) ermittelt.

Das Ergebnis zeigt, daß innerhalb des Intervalls $[\alpha_E, \alpha_p]$ Werte für α existieren, für welche die Approximation nach beiden Fehlerfunktionalen wesentlich besser ist als mit den aus (18) und (22) berechneten Werten. Außerdem hat sich erwiesen, daß mit α_E die Extremwerte und mit α_p die Nullstellen von $K(t)$ gut approximiert werden. Für kleines d folgt aus (22) und (26) $\alpha_p \approx \omega_0$.

Diese Aussagen bestätigen, daß bei der Approximation von Impulsantwortfunktionen durch den Time-Scale-Faktor nicht nur das Abklingverhalten (Dämpfung) sondern, auch die Nullstellen (Eigenfrequenzen) bestimmt werden, was auch aus (10) ersichtlich wird. Es erscheint daher sinnvoll, die Wahl dieses Parameters durch ein Intervall zu begrenzen, dessen Randwerte aus den Formeln (18) und (22) berechnet werden.

3. Eine Methode zur Bestimmung der Impulsantwortfunktion aus der Wiener-Hopf-Integralgleichung

In [1] wird eine Modifizierung der Formel (22) angegeben, in der nur bekannte Größen zur Bestimmung von α verwendet werden. Dazu wird vorausgesetzt, daß $\Phi(t)$ Lösung der Differentialgleichung

$$D\Phi = \delta \tag{29}$$

ist (D stellt einen Differentialoperator, δ die Dirac-Funktion dar). Damit folgt unter Berücksichtigung von (1)

$$K = D\Psi, \tag{30}$$

und (22) nimmt eine für das gegebene Problem anwendbare Form an:

$$\alpha_p = \frac{||t^{1/2} D\dot{\Psi}||_{L^2[0,\infty)}}{||t^{1/2} D\Psi||_{L^2[0,\infty]}}. \tag{31}$$

In diesem Fall braucht jedoch die Formel (31) nicht mehr angewendet zu werden, da die gesuchte Impulsantwortfunktion unmittelbar aus der Beziehung (30) berechnet werden kann.

Praktische Schwierigkeiten bildet die Bestimmung des Differentialoperators D. Dieses Problem kann z. B. durch die Konstruktion eines geeigneten Formfilters für das Eingangssignal gelöst werden.

Im Beispiel (24), (25) mit exponentiell korrellierter Eingangsgröße lautet dieser Operator

$$D\left(\frac{d}{dt}\right) = -\frac{1}{2a}\frac{d^2}{dt^2} + \frac{a}{2}. \qquad (32)$$

Man kann sich leicht davon überzeugen, daß durch die Anwendung von (32) und (30) die Impulsantwortfunktion (26) berechnet wird.

Literatur

1 BARS, R., On some questions of linear system identification by the Laguerre orthonormal system, Periodica Polytechnica, Electrical Engineering **19**, No. 2, (1975), 103–112.
2 BARS, R.; MAYER, I., Algorithmus for evaluating the Laguerre and X-expansion coefficients of transfer functions, Problems of Control and Information Theory **6** (3), (1977), 249–260.
3 PARKS, T. W., Choise of time scale in Laguerre approximations using signal measurements, IEEE Trans. Automat. Control. Oct. 1971, 511–513.
4 HEIMANN, B.; UECKERDT, R., Bestimmung der Parameter in linearen zeitinvarianten stochastisch erregten Systemen, Materialien des Problemseminars „Stochastische Schwingungen und Identifikation", TU Dresden, Oktober 1978.
5 MÜLLER, P. C.; SCHIEHLEN, W. O., Lineare Schwingungen, Akademische Verlagsgesellschaft, Wiesbaden 1976.

Eingereicht am 25. 5. 1981

Anschrift: Dipl.-Math. W. TISCHER, Inst. f. Mechanik der AdW der DDR, Rudower Chaussee 5, DDR-1199 Berlin-Adlershof

ZAMM 62, 276 (1982)

H. HEINRICH

Bemerkung zu „M. Becker / W. Hauger. Ein sehr einfaches Verfahren zur Bestimmung der Wurzeln algebraischer Gleichungen dritten und vierten Grades".

ZAMM **61**-5 (1981), T 267–T 269 (siehe auch [3])

Dieser auf der GAMM-Tagung 1980 gehaltene Vortrag gibt Anlaß, auf andere Verfahren zur numerischen Lösung biquadratischer Gleichungen

$$f(x) = x^4 + a_3x^3 + a_2x^2 + a_1x + a_0 = 0 \qquad (1)$$

hinzuweisen, die seit langem bekannt und insofern einfacher sind als das jetzt von den Autoren vorgestellte, als sie mit der Bestimmung einer reellen Lösung einer Gleichung dritten (statt sechsten) Grades

$$R(z) = z^3 + \alpha z^2 + \beta z + \gamma = 0 \qquad (2)$$

auskommen. Man vergleiche dazu die schon 1947 von G. OPITZ gegebene Übersicht [1].

Der in dem Vortrag unter (7) und (8) zu findende Ansatz

$$x_{1,2} = \varepsilon_1 \pm \sqrt{-\eta_1}, \qquad x_{3,4} = \varepsilon_2 \pm \sqrt{-\eta_2} \qquad (3)$$

für die Lösungen der Gleichung (1) läßt die Verwandtschaft mit dem von mir 1941 in [2] veröffentlichten Verfahren erkennen, das in [1] unter der Nummer 3 läuft. Bei ihm wird das Polynom $f(x)$ mittels des HORNERschen Schemas nach Potenzen von

$$t = x + a_3/4 \qquad (4)$$

entwickelt, die Gleichung (1) dadurch in

$$g(t) := f(t - a_3/4) = t^4 + at^2 + bt + c = 0 \qquad (5)$$

umgewandelt und t in der Form

$$t = u \pm \sqrt{-v} \qquad (6)$$

angesetzt. Die Koordinatenverschiebung erweist sich als die entscheidende Maßnahme, die die Resolvente $R(z)$ auf den Grad 3 reduziert. Ihre Koeffizienten hängen in einfachster Weise von den Koeffizienten von $g(t)$ ab:

$$\alpha = 2a, \qquad \beta = a^2 - 4c, \qquad \gamma = -b^2. \qquad (7)$$

Sie hat wegen $\gamma < 0$ (mindestens) eine positive Nullstelle z^*. (Der Fall $b = 0$ führt auf eine quadratische Gleichung (5) und ist daher uninteressant.) Ist z^* numerisch bestimmt, so erhält man über die Werte

$$u_{1,2} = \pm\frac{1}{2}\sqrt{z^*},$$

$$v_k = g'(u_k)/4u_k = u_k^2 + a/2 + b/4u_k, \qquad k = 1, 2, \qquad (8)$$

die gesuchten Lösungen der gegebenen Gleichung (1):

$$x_{1,2} = u_1 - \frac{1}{4}a_3 \pm \sqrt{-v_1},$$
$$x_{3,4} = u_2 - \frac{1}{4}a_3 \pm \sqrt{-v_2}. \qquad (9)$$

Literatur

1 OPITZ, G., Praktische Verfahren zur Lösung von Gleichungen vierten Grades, ZAMM **25/27** (1947), 171–173.
2 HEINRICH, H., Zur rechnerischen Auflösung einer Gleichung vierten Grades, ZAMM **21** (1941), 304–307.
3 BECKER, M.; HAUGER, W., Eine sehr einfache Methode zur Bestimmung der Wurzeln algebraischer Gleichungen 3. und 4. Grades, Forschungs- u. Seminarbericht des Lehr- u. Forschungsgebietes Bautechnik, Univ.-GH-Wuppertal; FB Bautechnik, Ber.-Nr. **1**/1981

Eingereicht: 21. 8. 1981

Anschrift: Prof. Dr. HELMUT HEINRICH, Friedrich-Hegel-Straße 31, DDR-8027 Dresden

BUCHBESPRECHUNGEN

Truesdell, C. III, The Tragicomical History of Thermodynamics, 1822—1854, Studies in the History of Mathematics and Physical Sciences 4, Berlin-Heidelberg-New York, Springer-Verlag 1980. XII, 372 S., 8 Abb., DM 99,—, US $ 58.50. ISBN 3-540-90403-4

Truesdell versteht Geschichte der exakten Naturwissenschaften als mathematische Entwicklung physikalischer Ideen. Indessen kongruiert der Gang der Historie nicht automatisch mit den logischen Explikationen. Ein solches Mißverhältnis kennzeichnet im 19. Jahrhundert insonderheit die Thermodynamik; denn (so der Tenor des vorliegenden Werkes) sie bleibt deutlich unter dem von Hydrodynamik, Elastomechanik und der Theorie des Elektromagnetismus bestimmten Niveau, was die Schärfe der Begriffsbildung, sowie die ernste Strenge und mathematische Brillanz der Beweisführung angeht. Mangel an Ernst, Können und Wissen, wenn er zu Verwicklungen führt, wirkt aber komisch, weshalb die hier erzählte Geschichte Züge von Komik trägt; der unfreiwilligen, versteht sich, welche meistens, zumal es ja um an sich durchaus ernst zu nehmende Dinge geht, des Tragischen nicht enbehrt, — eben weil es tragisch ist, wenn etwas Bedeutsames nicht so getan wird, wie es rechtens geschehen müßte. Da also nur im Kontrast zu einem Ideal die Entwicklung der Thermodynamik von Laplace bis Clausius als Tragikomödie erscheinen kann, bedeutet diese zu schreiben zuerst, die fundamentalen Gedanken der Pioniere der fraglichen Disziplin in logisch-mathematisch idealer Weise zu entfalten. Diese Aufgabe bewältigt der Verf. glänzend, ja, mit einer solchen Bravour, daß er dabei jene alte Wissenschaft von der Wärme noch um einige bislang unbekannte Theoreme bereichert. Aber sollte nun der Ernst des Mathematikers dieses vornehme Verdienst um den Preis der Treue errungen haben, die der Historiker den wirklichen Taten seiner Helden schuldet? Nein, eine derartige Befürchtung, wiewohl naheliegend, erweist sich als unbegründet. Truesdell hat Gleichung um Gleichung der originalen Arbeiten nachgerechnet, Argument um Argument durchdacht. Freilich hat er dabei die Gedanken tiefer als ihre Urheber selbst verstanden und lichtvoller dargestellt. Aber daß trotzdem der Verbraucher (natürlich des eigenen Quellenstudiums nicht enthoben) erleben kann, wie die tragikomische Geschichte der Thermodynamik tatsächlich vonstatten ging, das qualifiziert das Buch zu einem Kunstwerk. Nicht zuletzt ist ja „tragikomisch" eine ästhetische Kategorie. Und da wir gerade darüber sprechen: Welche Machtfülle aber auch Geschmeidigkeit offenbart die englische Sprache in Truesdell's Handschrift!

Dem eigentlichen Schauspiel vorangestellt ist ein Prolog, in dem Laplace, Biot und Poisson auftreten. Die beiden ersten Akte sind zwei komplementären Ideen gewidmet: Dissipation ohne Arbeit und Arbeitsleistung ohne Dissipation. Der Schöpfer des einen chimärischen Konzepts war Fourier, der des anderen Carnot. Dieser junge Mann ist der Genius der klassischen Thermodynamik und zugleich der Vater ihrer abstrusen Eigenwilligkeiten. Deren erstaunlichste (obschon sie aus alter Gewohnheit wie selbstverständlich hingenommen wird) ist die folgende: Während alle anderen physikalischen Disziplinen auf fundamentalen, allgemeinen Aussagen über die wirkliche Beschaffenheit der Materie beruhen, besagt das grundlegende Axiom der Thermodynamik, wie die Natur *nicht* funktioniert! (Nichtexistenz von perpetua mobilia.).

Nach einem kurzen, von Clapeyron und Duhamel bestrittenen Zwischenspiel handelt der dritte Akt des Dramas von der Suche nach dem Energiekonzept (Mayer, Holtzmann, Helmholtz, der frühe Joule), dessen rationale thermodynamische Variante, der Begriff der inneren Energie, dann von dem zweiten überragenden Helden unserer Geschichte gefunden wird, von Clausius. Er hat im vierten Akt seinen ersten Auftritt, wo uns auch Rankine begegnet, eine frühe Version des Entropiekonzepts ohne Echo vortragend. Und nachdem in einem erneuten, langen Zwischenspiel vornehmlich die Experimentalphysik hinreichend Verwirrung gestiftet hat, wird „in einem dunklen und leeren Theater" vollends ein Gegenplan (antiplot) entwickelt. Ferdinand Reech, ein völlig unbekannter Akteur, entwirft als fünften Akt eine viel zu allgemeine Theorie, ohne sie aufs rechte Maß reduzieren zu können. Es ist ein „schismatischer" Entwurf; der fünfte Akt muß im orthodoxen Stil noch einmal geschrieben werden. Das schließlich bewerkstelligen Kelvin und Clausius, indem sie die absolute Temperatur und den zweiten Hauptsatz etablieren. Als Überschrift für den Epilog findet Truesdell nur ein deutsches Wort: Götterdämmerung.

Dresden H.-G. Schöpf

Ebert, F., Strömung nicht-newtonscher Medien. Braunschweig, Vieweg 1980. X, 190 S., DM 29,80. ISBN 3-528-03339-8

Fluide, nach dem Aggregatzustand in Flüssigkeiten und Gase unterschieden, sind definitionsgemäß solche Stoffe, die jeder noch so kleinen Scherbeanspruchung durch Bewegung nachgeben. Dabei werden auch Medien mit Fließgrenze wie Zahnpasta und Schokoladengrundmasse unter die Fluide gerechnet, wenn man sich für ihr Strömungsverhalten interessiert. Die newtonschen Fluide wie z. B. Wasser und Luft sind dadurch ausgezeichnet, daß zwischen Schubspannung τ und Geschwindigkeitsgradient ein linearer Zusammenhang $\tau = \eta\dot{\gamma}$ besteht, wobei η eine Materialkonstante ist.

Blut, Gelenkflüssigkeiten; Polymerschmelzen, Farben, Lacke u. s.w. sind Beispiele von nichtnewtonschen Flüssigkeiten, bei denen η von $\dot{\gamma}$ und gegebenenfalls von der Zeit t abhängt. Der vorliegende Text ist den Bewegungen nichtnewtonscher Flüssigkeiten gewidmet. Er ist als Einführung für den Ingenieur, der über Kenntnisse in der klassischen Strömungslehre verfügt, gedacht. Im Rahmen einer Diskussion der Funktion $\eta(\dot{\gamma}, t)$ werden die Flüssigkeiten nach Ihrem Fließverhalten klassifiziert. Die experimentelle Bestimmung der Fließeigenschaften wird mit dem Schwerpunkt ihrer technischen Realisierung erörtert. Die Fließgesetze werden mit Hilfe der Tensorrechnung formuliert. Ausführlich werden die Strömungen in Viskosimetern behandelt. Ein eigenes Kapitel ist der Rohrhydraulik gewidmet. Die Darstellung beschränkt sich auf laminare Strömungen. Der Anhang zur Tensorrechnung begnügt sich mit der Wiedergabe von Tatsachen, die in einschlägigen Büchern verständnisfördernder nachgelesen werden können.

Auch Spezialisten der physikalischen Chemie und der Kontinuumsmechanik, die sich mit rheologischen Flüssigkeiten beschäftigen, kann das vorliegende Buch empfohlen werden, wenn sie daran interessiert sind zu erfahren, wie der Strömungstechniker dieses Gebiet betrachtet.

Berlin J. Förste

Parkus, H. (ed.), Electromagnetic Interactions in Elastic Solids. CISM Courses and Lectures 257. Wien-New York, Springer-Verlag 1979. VI, 425 S., 15 Abb., S 486,—. DM 68,—. US $ 38.10. ISBN 3-211-81509-0

Der von dem bekannten Experten der Festkörpermechanik H. Parkus (Wien) herausgegebene Bd. 257 der am Internationalen Mechanik-Zentrum in Udine (Italien) gehaltenen Vorlesungen beschäftigt sich mit zahlreichen Aspekten der elektromagnetischen Wechselwirkungen in elastischen Festkörpern. 6 international anerkannte Wissenschaftler gehen ausführlich auf Fragen des mechanisch-thermisch-elektromagnetischen Feldkopplungs-Problems ein. J. B. Alblas behandelt die allgemeinen Grundlagen der Theorie der Elektro- und Magnetoelastizität. W. Nowacki, der vor allem durch seine grundlegenden Arbeiten zur Theorie der mikropolaren Medien bekannt wurde, untersucht Fragen der Piezoelektrizität und der Magnetoelastizität. K. Hutter nennt seinen Beitrag „Thermodynamische Aspekte der Feld-Materie-Wechselwirkungen". Die Kontinuumsthermodynamik ist jedoch die Grundlage aller Beiträge der Autoren dieses Bandes. G. A. Maugin beschäftigt sich mit der klassischen Magnetoelastizität und den defektbehafteten Ferromagnetika. Interessant ist auch der Beitrag von A. Prechtl über einige relativistische Aspekte der elektromagnetischen Wechselwirkungen in elastischen Festkörpern. Der Herausgeber, H. Parkus, gibt neben der Darstellung der Grundlagen auch konkrete Anwendungsbeispiele (z. B. magnetoelastische Platte) an. Nach Ansicht des Rezensenten ist dieses Buch ein grundlegendes Werk der modernen Kontinuumsmechanik mit perspektivischer Zielrichtung, von der noch wichtige Ergebnisse zu erwarten sind. Es ist überdies eines der ersten Bücher seiner Art, dem sicher noch weitere folgen dürften. Dem auf dem Gebiet der Festkörpermechanik und verschiedenen Disziplinen der Physik (Festkörperphysik) tätigen

Forscher ist es ebenso zu empfehlen wie den Kollegen der Mathematik (insbesondere der Analysis partieller Differentialgleichungssysteme). Als Lehrbuch dürfte es in höheren Semestern an Universitäten mathematisch-naturwissenschaftlicher Fachrichtungen sowie in der Mechanik u. U. ebenfalls einsetzbar sein. Jedoch erschwert die unterschiedliche Symbolik und Begriffsbildung der einzelnen Autoren einen derartigen Einsatz. In mathematischen und naturwissenschaftlich-technischen Fachbibliotheken sollte dieses Werk erhältlich sein.

Karl-Marx-Stadt B. MICHEL

Flügge, S., Mathematische Methoden der Physik II, Geometrie und Algebra. Berlin-Heidelberg-New York, Springer-Verlag 1980. VI, 174 S., 19 Abb., DM 38,–. US $ 22.50. ISBN 3-540-10062-8

In dem vorliegenden Band werden geometrische und algebraische Methoden der mathematischen Physik dargestellt. Zunächst erfolgt eine Einführung in die elementare Vektor- und Tensoranalysis: Gradient, Divergenz, Rotation, Wirbel, Quellen, Tensoren, Integralsätze. Im 2. Kapitel wird die RIEMANNsche Geometrie behandelt: Vektor- und Tensoranalysis, metrischer Tensor, Krümmungstensor, Integrabilität, Variationsprinzip und Feldgleichungen der Relativitätstheorie. Das 3. Kapitel beschreibt algebraische Methoden der mathematischen Physik. Es werden eingeführt: Körper, Ringe, Gruppen, Darstellungen endlicher Gruppen, Quaternionen, Spinoren, PAULImatrizen. Als Beispiel wird die Permutationsgruppe des Austauschs von drei Teilchen betrachtet. Ferner werden beschrieben die Darstellungen der Gruppen SU2, SU3, SU4 und Verallgemeinerungen hierzu. Für höherdimensionale Darstellungen der SU3 wird der konstruktive Aufbau von Multipletts mittels Verschiebeoperatoren behandelt und ein Verfahren zur Bestimmung der Multiplizität beschrieben. Zu jedem Kapitel gehört eine Sammlung von Aufgaben mit Lösungen. Für das Verständnis sind erforderlich die elementaren Grundkenntnisse der analytischen Geometrie und der mathematischen Analysis. Der Leser erhält durch dieses Buch geometrische und algebraische Methoden vermittelt, die für die Physik wichtig sind. Es ist als Hochschullehrbuch für Studenten zu empfehlen.

Berlin E. HEYN

Ciarlet, G. / Rabier, P., Les Equations de von Kármán. Lecture Notes in Mathematics 826. Berlin-Heidelberg-New York, Springer-Verlag 1980. VI, 181 S., 15 Abb. DM 21,50. US $ 12.70. ISBN 3-540-10248-5

Die Autoren untersuchen im Rahmen einer geometrisch nichtlinearen Theorie das statische Gleichgewicht einer dünnen Platte aus linear elastischem Material, die durch in der horizontalen Mittelebene der Platte und in der dazu senkrechten Richtung wirkende Kräfte belastet wird. Ihr wesentliches Anliegen besteht in einer konsequenten Herleitung zweidimensionaler Plattengleichungen aus den dreidimensionalen Grundgleichungen der geometrisch nichtlinearen Elastizitätstheorie sowie in einer modernen mathematischen Behandlung der so gewonnenen Gleichungen.

Ausgangspunkt der Untersuchungen ist ein dreidimensionales Randwertproblem, das die eingangs genannte Situation beschreibt. Lösungen dieser Aufgabe werden in Form einer Reihe nach Potenzen der kleinen Plattendicke gesucht. Im Sinne einer asymptotischen Entwicklung beschränken sich die Autoren auf das erste Glied dieser Reihe. Sie zeigen, daß dieses Vorgehen unter gewissen Regularitätsannahmen der KIRCHHOFF-LOVEschen Hypothese entspricht und auf zweidimensionale, in den (horizontalen und vertikalen) Verschiebungen formulierte Plattengleichungen führt. Sie zeigen ferner, daß die so gewonnenen Gleichungen unter gewissen Regularitätsannahmen zu den zweidimensionalen von KÁRMÁNschen Gleichungen äquivalent sind, die mit Hilfe der Spannungsfunktion und der vertikalen Verschiebung formuliert werden. Auf diese Weise erfahren die von KÁRMÁNschen Gleichungen eine asymptotische Begründung.

Anschließend wird bewiesen, daß bei hinreichend glatten Kräften jedes der beiden zweidimensionalen Plattenprobleme lösbar ist, wobei alle Lösungen so glatt sind, daß die oben genannte Äquivalenzaussage gilt. Der weiteren Untersuchung können also die von KÁRMÁNschen Gleichungen zu Grunde gelegt werden.

Bei speziellen Annahmen über die Kräfte lauten die von KÁRMÁNschen Gleichungen (in kanonischer Form)

$$u \in H: u - \lambda L u + C(u) - \delta F = 0\,, \qquad (x)$$

wobei H ein Hilbertraum, $F \in H$ fixiert, L ein linearer, selbstadjungierter, vollstetiger Operator und C ein vollstetiger 3-Potenzoperator ist; $(u, \lambda, \delta) \in H \times R \times R$ bezeichnet die vertikale Verschiebung und die Größe der horizontalen bzw. vertikalen Kraft. Gesucht ist insbesondere die Anzahl aller kleinen Lösungen von (x) für (λ, δ)-Werte in einer Umgebung von $(\lambda_0, 0)$, wobei λ_0 eine charakteristische Zahl von L ist. Derartige Fragestellungen können mit Methoden der analytischen Verzweigungstheorie behandelt werden. Die Autoren bringen hier eine vollständige und übersichtliche Darstellung dieser Methoden für den Fall, daß λ_0 eine einfache charakteristische Zahl von L ist. Die erzielten Ergebnisse werden abschließend am Beispiel der Kreisplatte kurz erläutert.

Das Buch bietet eine klar und übersichtlich geschriebene Einführung in die nichtlineare Plattentheorie und die dort verwendeten mathematischen Methoden; benötigte Hilfsmittel aus der Theorie der SOBOLEWräume werden in einem Anhang zusammengestellt. Deshalb kann das Buch all denen empfohlen werden, die sich in die genannte Problematik einarbeiten wollen.

Berlin R. HÜNLICH

Doebner, H. D. (ed.), Differential Geometric Methods in Mathematical Physics. Proceedings, Clausthal, 1978. Lecture Notes in Physics 139. Berlin-Heidelberg-New York, Springer-Verlag 1981. VII, 329 S., DM 38,–. US $ 18.10. ISBN 3-540-10578-6

Nachdem von EINSTEIN der fundamentale Wert der RIEMANN-Geometrie für die Gravitationsphysik entdeckt worden war, bahnte sich eine starke interdisziplinäre Verflechtung von Feldtheorie und Differentialgeometrie an (z. B. WEYL-Geometrie, FINSLER-Geometrie, KÄHLER-Geometrie). Im Zusammenhang mit der seit etwa einem Jahrzehnt intensiv beschriebenen neuen theoretisch-physikalischen Forschungsrichtung, die man durch die Stichworte Eichfeldtheorien und Supergravitation charakterisieren kann, erlebt diese Kooperation einen Höhepunkt. Die vorliegenden Proceedings der in Clausthal im Juli 1978 veranstalteten internationalen Konferenz, die etwa 40 Spezialisten aus aller Welt zu einem regen Meinungsaustausch zusammenführte, widerspiegeln die Brennpunkte der heute aktuellen Forschung auf diesem Gebiet: Quantisierungsmethoden und spezielle Quantensysteme (geometrische Quantisierung, Vektorfeld-Quantisierung, Quantisierung von stochastischen Phasenräumen, Dynamik magnetischer Monopole), Eichfeldtheorien (Phasenraum der klassischen YANG-MILLS-Gleichung, Eich-Geometrodynamik, exzeptionale Eichgruppen), elliptische Operatoren und Spektraltheorie (ATIYAH-SINGER-Theorem in der Quantenfeldtheorie, Anwendung der Spektraltheorie auf Phasenübergänge), geometrische Methoden und globale Analysis (Systeme auf Nicht-HAUSDORFF-Räumen und auf nichteuklidischen Räumen, WEYL-Geometrie, LORENTZ-Mannigfaltigkeiten, Einbettungs-Mannigfaltigkeiten).

Meines Erachtens kommt dem vorliegenden Buch dadurch ein besonderes Verdienst zu, daß es eine gemeinsame Sprache zwischen Mathematikern und Physikern findet und damit interdisziplinär stark befruchtend wirkt. Gerade aus diesem Grund möchte ich es insbesondere der jüngeren Generation beider Fachrichtungen empfehlen, sofern die Forschungsziele korrespondieren.

Jena E. SCHMUTZER

Blaquiere, A. / Fer, F. / Marzollo, A. (eds.), Dynamical Systems and Microphysics. CISM Courses and Lectures 261. Wien-New York, Springer-Verlag 1980. IX, 412 S., 33 Abb., DM 66,–. S 472,–. US $ 39.00. ISBN 3-211-81533-3

Der vorliegende Band der international bekannten Buchreihe von Vorträgen am Internationalen Mechanik-Zentrum in Udine, Italien, (CISM) umfaßt eine Sammlung von Arbeiten von etwa 30 Autoren aus 9 Ländern Westeuropas und der USA zur Thematik „Dynamische Systeme und Mikrophysik“. Das Anliegen des Buches wird von den Herausgebern klar umrissen: Anwendung neuerer mathematischer Methoden der Systemtheorie auf aktuelle Fragen der modernen Physik. Der Band enthält das Ergebnis einer Vortrags- und Diskussionsrunde am Centre for Mechanical Science (CISM) im Jahre 1979 in Udine. Die Beiträge des Teils 1 befassen sich mit Fragen der Optimierungstheorie und der Stochastik von Quantenprozessen. Der Teil 2 ist Problemen der Evolution sowie der Instabilität gewidmet, während der Teil 3 des Buches solche grundlegenden Probleme der Quantenmechanik wie „Indeterminismus“ und „Nichtseparabilität“ berührt. Die einzelnen Beiträge sind von kompetenten Fachleuten geschrieben worden. Die Sammlung gibt einen guten Überblick in allerdings sehr kurzen Einzelbeiträgen über aktuelle Fragen der Mikrophysik, ist aber für den Nicht-Spezialisten manchmal schwer lesbar. Dennoch sollte das Werk einen großen Leserkreis finden, da es „wirklich das erste Werk ist, das als Ergebnis eines Treffens so vieler Spezialisten aus verschiedenen Gebieten der Physik und Mathematik in dem Bemühen entstand, gemeinsam einer Lösung der Probleme und Schwierigkeiten der gegenwärtigen Mikrophysik näherzukommen.“, wie der international bekannte französiche Physiker und Nobelpreisträger LOUIS DE BROGLIE in seiner Vorbemerkung zu dem Werk treffend feststellte. Der besondere Reiz des Bandes, der für Physiker und Mathematiker gleichermaßen von Interesse sein dürfte, liegt gerade in seiner hohen Aktualität, die durch manchmal kontroverse Auffassungen der einzelnen Autoren nur unterstrichen wird. Die z. T. ausführlichen Quellenverzeichnisse der Einzelbeiträge sind sicher für die intensivere Beschäftigung mit diesen Fragen von großem Wert.

Karl-Marx-Stadt B. MICHEL

Jaffe, A. / Taubes, C., Vortices and Monopoles, Structure of Static Gauge Theories, Progress in Physics 2, Boston-Basel-Stuttgart, Birkhäuser Verlag 1980, 287 S., DM 30,—. ISBN 3-7643-3025-2

Die während der letzten zehn Jahre gewonnenen experimentellen Ergebnisse der Elementarteilchenphysik haben eine Fülle von Hinweisen darauf erbracht, daß es Eichtheorien sind, die eine adäquate theoretische Beschreibung dieser Phänomene gestatten. In der vorliegenden Monographie geht es jedoch um klassische, d. h. nichtquantisierte Eichtheorien, und insbesondere um die YANG-MILLS-HIGGS (YMH)-Theorien, wo die Eichfelder an das skalare HIGGS-Feld koppeln, das seinerseits noch einer charakteristischen Selbstwechselwirkung („Flaschenpotential“) unterliegt. Damit wird auch die schon seit langem diskutierte phänomenologische LANDAU-GINZBURG-Theorie der Supraleitung erfaßt. Gegenstand der Untersuchung sind statische Lösungen, die hier als Lösungen über einem euklidischen R^d mit um eins verminderter Dimension behandelt werden, und die überdies durch einen endlichen Wert des Wirkungsintegrals ausgezeichnet sind. In dieser euklidischen Formulierung sind es dann nur die Fälle $d = 2$, 3 und 4, für die es überhaupt nichttriviale Lösungen gibt. Das entspricht den Vortices der LG-Theorie (falls auf die abelsche Gruppe U(1) spezialisiert wird), den Monopolen der nichtabelschen YMH-Theorien und den Instantonen der reinen YM-Theorien. Als wichtigste Resultate hinsichtlich der Vortices und Monopole seien genannt: Jede glatte, d. h. zur Klasse C^∞ gehöhrende Lösung mit endlicher Wirkung ist durch Homotopieklassen, und damit durch gewisse ganze Zahlen gekennzeichnet. Das sind insbesondere die Vortexzahl, die den Magnetfluß angibt, bzw. die Monopolzahl im Falle der SU(2)-Monopole. Für spezielle Werte der HIGGS-Kopplungskonstante wird die Existenz von Lösungen zu beliebigen ganzzahligen Werten der Vortex- bzw. Monopolzahl bewiesen; jedoch läßt sich aus dem Beweis keine explizite Konstruktionsvorschrift ableiten, wie sie etwa für die Instantonen bekannt ist. Für alle glatten Lösungen endlicher Wirkung werden die Abfallseigenschaften im Unendlichen bestimmt, die für die Monopole mit Massenparametern der YMH-Theorie in Verbindung stehen. Es sollte aber in diesem Zusammenhang daran erinnert werden, daß die Rolle von klassischen Monopollösungen im Rahmen einer Quantenfeldtheorie, und damit auch die physikalische Relevanz solcher Massenparameter noch keineswegs geklärt sind.

Die Monographie verdient zweifellos das größte Interesse von Mathematikern und Physikern, die sich mit Eichfeldtheorien befassen; im Hinblick auf die angewandten Methoden (Funktionalanalysis und Topologie) ist es aber auch allen denjenigen zu empfehlen, die auf dem weiteren Gebiet der Solitonen arbeiten.

Berlin E. WIECZOREK

Eckhaus, W. / van Harten, A., The Inverse Scattering Transformation and the Theory of Solitons. An Introduction. North-Holland Mathematics Studies 50. Amsterdam-New York, North-Holland Publ. Co. 1981. XI, 224 S., US $ 31.75. Dfl. 65.00. ISBN 0-444-86166-1

This book provides an introduction to inverse scattering transformations and their connection to the theory of solutions. The way of presenting the theory is understandable and acceptable, mathematically satisfactory and rigorous. Some sophisticated proofs are shortened. The reader is assumed to be uninitiated and familiar only with some fundamental topics of functional analysis.

The content follows the historical and chronological line of development of inverse scattering theory. In chapter 1 the KORTEWEG-DE-VRIES (KdV) equation is introduced. The existence and uniqueness of solutions of K.d.V. and modified K.d.V. equations are proved. Chapter 2 summarizes the results of the GARDNER-GREENE-KRUSHAL-MIURA method for finding the solution of K.d.V. equation by means of inverse scattering transformations. This method is based on the inverse scattering for SCHRÖDINGER equations with timedependent potentials (see also chapter 4). In chapter 3 the more general problem is studied whether there are other equations, than K.d.V. equation, the solutions of which are isospectral potentials for the SCHRÖDINGER operator. The LAX approach to this problem is given, using the concept of unitary equivalence between selfadjoint operators in HILBERTspaces, and studied in some examples. Moreover, the LAX formalism is extended to nonselfadjoint operators. Chapter 4 and 5 are selfcontained and treat extensively the direct and inverse scattering and spectral theory for SCHRÖDINGER equations and generalized ZAKHAREV-SHABAT systems, respectively. In chapter 4 the spectral theory for SCHRÖDINGER operators in $L^2(R^1)$ is described. Results are given with respect to regularity, dependence of data, and asymptotic behaviour of the solutions of the SCHRÖDINGER equation. Their FOURIER integral expressions are given and the GELFAND-LEVITAN-MARCHENKO integral equation is derived. Chapter 5 describes the same theory for the generalized ZAKHAROV-SHABAT system and is organized in a similar way as chapter 4. Finally, chapter 6 gives some applications of the inverse scattering transformation theory in order to show merely how the method works. In chapter 7 solutions of perturbed integrable evolution equations are considered. In the framework of inverse scattering transformations it is needed the time evolution of the scattering data. The evolution equations for the spectral data are derived in the case of the SCHRÖDINGER equation corresponding to a perturbed K.d.V. equation and for generalized ZAKHAROV-SHABAT systems.

Berlin M. DEMUTH

Hawking, S. W. / Roček, M. (eds.), Superspace and Supergravity. Cambridge, Cambridge University Press 1981. XII, 527 S., £ 24.00 A H/C. ISBN 0-521-23908-7

Nachdem das EINSTEINsche Programm einer einheitlichen Feldtheorie der Materie über ein halbes Jahrhundert ohne durchschlagenden Erfolg geblieben war, setzte vor etwa einem Jahrzehnt sehr intensiv eine neue Forschungsrichtung mit derselben Zielstellung ein: Die Entdeckung der Supersymmetrie für Bosonen (ganzzahliger Spin) und Fermionen (halbzahliger Spin) führte zur Einführung des „Superspace“ als der zugehörigen geometrischen Mannigfaltigkeit. Falls die physikalische Idee der Quantisierung der Gravitation richtig ist, also Gravitonen (Bosonen mit Spin 2) existieren, so ist in diesem erweiterten geometrischen Rahmen auch Platz für die Gravitation geschaffen. Man kommt dann zu einer quantisierten Neufassung der Gravitation, nämlich zur Supergravitation mit ihren Implikationen (Gravitino als Spin 3/2-Teilchen). Ein großer Teil der theoretischen Ansätze in dieser Richtung geht heute von einem 8-dimensionalen Superraum aus, der durch die 4-dimensionale Raum-Zeit (kommutative Koordinaten) und vier GRASSMANN-Dimensionen (antikommutierende Koordinaten) aufgespannt wird.

Das hier kurz umrissene Gedankengebäude bildet den Inhalt des vorliegenden Buches, das als Proceedings die beim Nuffield-Workshop in Cambridge (England) vom 16. 6.–12. 7. 80 gehaltenen Vorträge wiedergibt. Dort hatte sich für fast einen Monat ein relativ kleiner Kreis von nur 60 Spezialisten versammelt, um die herangereifte Thematik im einzelnen auszudiskutieren. Schwerpunkte der etwa 30 Beiträge waren dabei folgende Gebiete: Grundlegung von Supersymmetrie, Superspace und Supergravitation, Quantisierung (Hintergrund-Superfeld, Graphen, Eichung, Regularisierung, Eigenzeit-Zugang), erweiterte Supergravitation (konforme Supergravitation, Konstruktion von Hilfsfeldern, 5-dimensionale Supergravitation, Darstellungsfragen, Gruppenzerlegungen, Superalgebren), 8-dimensionale Supergravitation (spontane Supersymmetrie-Brechung, kosmologische Konstante, Grand Unification), KÄHLER-Räume mit Supersymmetrie (KÄHLER-Geometrie, Eigenfunktionen in KÄHLER-EINSTEIN-Räumen) sowie Instabilitäten, Gittertheorie etc.

Für den auf diesem neuen Gebiet forschenden Theoretischen Physiker und Mathematiker stellt dieser Sammelband ein unentbehrliches Hilfsmittel dar.

Jena E. SCHMUTZER

Krylov, N. V., Controlled Diffusion Processes, Applications of Mathematics 14, Berlin-Heidelberg-New York, Springer-Verlag 1980, XII, 308 S., DM 79,–. US $ 44.30. ISBN 3-540-90461-1. (Translation from the Russian)

Die Theorie der stochastischen Steuerung (stochastische Kontrolltheorie) ist eine relativ junge mathematische Disziplin. Während ihrer Existenz erschienen zahlreiche Arbeiten zum Fall des quadratischen Gütefunktionals (diskrete stochastische Programmierung; lineare Systeme und quadratische Gütefunktionen). Der Gegenstand des vorliegenden Buches ist die Steuerungstheorie für Diffusionsprozesse, d. h. hier — die Steuerungstheorie stochastischer Differentialgleichungen in ihrer integralen Form mit stochastischem ITO-Integral. Vorausgesetzt wird beim Leser die Bekanntschaft mit den Grundzügen der stochastischen Integration. Grundlagen dieser Theorie sind zur Erleichterung in einem Anhang zusammengestellt. Bezüglich der Theorie der Differentialgleichungen werden demgegenüber kaum Kenntnisse vorausgesetzt. Die weiter oben genannten Gebiete und die deterministische Steuerungstheorie werden nicht betrachtet.

Das allgemeine Anliegen des Buches besteht im Beweis der Richtigkeit der BELLMANNschen Differentialgleichung für die Kostenfunktion und in der Aufstellung der Regeln zur Konstruktion ε-optimaler (näherungsweiser) Steuerstrategien für die gesteuerten Diffusionsprozesse. Hervorzuheben sind die Behandlung zeitoptimaler Probleme und der eindimensionalen kontrollierbaren Diffusion, die L_p-Abschätzungen für stochastische Integaldistributionen, Existenzaussagen für stochastische Gleichungen, die ITO-Formel für Funktionen, das BELLMANNsche Prinzip und die BELLMANNsche Gleichung, auch in ihrer normierten Form. Das Buch besteht aus 6 Kapiteln: 1. Einführung in die Theorie der gesteuerten Diffusionsprozesse, 2. Weitere Aussagen, 3. Allgemeine Eigenschaften der Gütefunktion, 4. Die BELLMANNsche Gleichung, 5. Zur Konstruktion ε-optimaler Strategien, 6. Gesteuerte Prozesse mit unbeschränkten Koeffizienten. Die normierte BELLMANNsche Gleichung. Die Bibliographie umfaßt ca. 75 Literaturstellen. Neben dem Vorwort den heuristischen Darlegungen in Kapitel 1 führen die Einleitungen zu den Kapiteln in die betreffende Problematik ein.

Kapitel 1 beeinhaltet allgemeine Konzepte, Methoden und Probleme der gesteuerten räumlichen Diffusion. Die rein heuristischen Darlegungen werden dann später unter speziellen Voraussetzungen exakt verifiziert. Mit Punkt 4 von Kapitel 1 beginnt die strenge mathematische Darlegung. Insbesondere wird auf zeitoptimale Probleme eindimensionaler gesteuerter Prozesse eingegangen. In Kapitel 2 werden u. a. stochastische Distributionen eingeführt, Existenz- und Einzigkeitsfragen bezüglich Lösungen stochastischer Differentialgleichungen beantwortet, und zwar für den Fall LIPSCHITZ-stetiger aber auch im Fall meßbarer Koeffizienten. Desweiteren werden Stetigkeits- und Differenzierbarkeitseigenschaften von Wahrscheinlichkeitsprozessen bezüglich eines Parameters nachgewiesen und die ITO-Formel für Funktionen mit verallgemeinerten Ableitungen in der Gestalt klassischer Funktionen bewiesen. In Kapitel 3 werden u. a. Stetigkeitseigenschaften der Gütefunktion unter minimalen Voraussetzungen an das System nachgewiesen und das BELLMANNsche Prinzip begründet. Anwendungen werden auf zeitoptimale Probleme vorgenommen. Kapitel 4 enthält den Nachweis von Glattheitseigenschaften der Gütefunktion, u. a. die Existenz der ersten Ableitung und Abschätzungen von oben und unten für die zweiten Ableitungen. Außerdem werden entartende gesteuerte Prozesse durch nichtentartende approximiert. Im nichtentartenden Fall wird die BELLMANNsche Gleichung unter der Voraussetzung der Existenz der verallgemeinerten Ableitung der Gütefunktion bewiesen. Kapitel 5 ist vor allem der Konstruktion der ε-optimalen MARKOVschen Strategien gewidmet, aus praktischen Gründen und ihrer Einfachheit wegen. Außerdem wird die BELLMANNsche Gleichung im entartenden Fall behandelt. Ausgehend von den sogenannten adjungierten MARKOVschen Strategien wird bewiesen, daß jede Lösung der BELLMANNschen Gleichung eine Gütefunktion ist.

Im Kapitel 6 werden die Ergebnisse der Kapitel 3-5 auf den Fall unbeschränkter Koeffizienten übertragen. Es werden Methoden zur Abschätzung der Ableitungen der Gütefunktion dargelegt, die normierte BELLMANNsche Gleichung behandelt und das asymptotische Verhalten der Gütefunktion in zeitoptimalen Problemen mit unbeschränkter Zeit für $T \to +\infty$ untersucht.

Berlin R. KLUGE

Everitt, W. N. (ed.), Ordinary and Partial Differential Equations, Proceedings, Dundee, Scotland 1978. Lecture Notes in Mathematics 827. Berlin-Heidelberg-New York, Springer-Verlag 1980. XVI, 271 S., 6 Abb., DM 29,–. US $ 17.20. ISBN 3-540-10252-3

Der vorliegende Berichtsband enthält die Plenarvorträge der fünften Konferenz über gewöhnliche und partielle Differentialgleichungen, die 1978 an der Universität in Dundee (Schottland) abgehalten wurde. Ehrenpräsident dieser Konferenz sollte Prof. ARTHUR ERDÉLYI sein, der aber Ende 1977 plötzlich verstarb. Die Konferenz wurde folglich dem Wirken Prof. ERDÉLYIS gewidmet, die Mehrzahl der Plenarvorträge hielten seine Schüler, Mitarbeiter oder Freunde. Im einzelnen enthält der Band folgende Beiträge: F. V. ATKINSON: Exponential behaviour of eigenfunctions and gaps in the essential spectrum; B. L. J. BRAAKSMA: LAPLACE integrals in singular differential and difference equations; DAVID COLTON: Continuation and reflection of solutions to parabolic partial difference equations; W. N. EVERITT: LEGENDRE polynomials and singular differential operators; GAETANO FICHERA: Singularities of 3-dimensional potential functions at the vertices and at the edges of the boundary; PATRICK HABETS: Singular perturbations of elliptic boundary value problems; F. A. HOWES and R. E. O'MALLEY Jr.: Singular perturbations of semilinear second order systems; H. W. KNOBLOCH: Higher order necessary conditions in optimal control theory; J. MAWHIN and M. WILLEM: Range of nonlinear perturbations of linear operators with an infinite dimensional kernel; ERHARD MEISTER: Some classes of integral and integro-differential equations of convolutional type; B. D. SLEEMAN: Multiparameter periodic differential equations; JET WIMP: Uniform scale functions and the asymptotic expansion of integrals.

Berlin K. R. SCHNEIDER

Doolan, E. P. / Miller, J. J. H. / Schilders, W. H. A., Uniform Numerical Methods for Problems with Initial and Boundary Layers. Dublin, Boole Press 1980. XII, 425 S., £ 28.00. US $ 60.00. ISBN 0-906783-01-1

Das Buch enthält zahlreiche bisher unveröffentlichte Ergebnisse der Autoren zu gleichmäßigen Differenzenverfahren für singulär gestörte Anfangs- und Randwertprobleme gewöhnlicher Differentialgleichungen. Der Terminus gleichmäßig wird dabei so verstanden, daß die Verfahren Fehlerabschätzungen gestatten mit einer Konstanten, die weder von der Schrittweite noch von einem in das Problem eingehenden kleinen Parameter abhängt. Es ist in die Abschnitte Anfangswertprobleme, Randwertprobleme und Numerische Ergebnisse unterteilt und enthält ein ausführliches Literaturverzeichnis.

Im ersten Abschnitt werden angepaßte numerische Verfahren für singulär gestörte lineare Anfangswertprobleme erster Ordnung diskutiert. Ausgehend von einer notwendigen Bedingung für den „angepaßten Faktor" (fitting factor) werden gleichmäßige Verfahren hergeleitet. Insbesondere wird dabei die für die Verkürzung der Rechenzeit wichtige Frage nach konstanten angepaßten Faktoren behandelt. Die Konstruktion gleichmäßiger Verfahren beliebiger Ordnung wird gezeigt und $O(h^2)$, $O(h^3)$ Verfahren werden explizit angegeben. Neben theoretisch abgesicherten Verfahren findet man auch Verfahren, von denen die Autoren annehmen, daß sie gleichmäßig konvergieren (z. B. ein Verfahren mit stückweise konstantem angepaßten Faktor für Systeme gewöhnlicher DGL 1. Ordnung).

Im zweiten Abschnitt werden angepaßte numerische Verfahren für singulär gestörte lineare Randwertprobleme 2. Ordnung untersucht. Dabei entarten die stetigen Probleme für gegen Null gehende Parameterwerte entweder auf Probleme 0. Ordnung oder auf Probleme 1. Ordnung mit nichtverschwindenden Koeffizienten bei der 1. Ableitung (d. h. Wendepunkte werden ausgeschlossen). Es werden gleichmäßige $O(h)$ konvergente Verfahren mit variablen angepaßten Faktoren angegeben und konstante Faktoren diskutiert. Die Beweise der gleichmäßigen Konvergenz stützen sich auf drei in der Literatur verwendete Techniken: der Zweigittermethode [IL'IN (1969), MILLER (1978)], der Ausnutzung asymptotischer Approximationen der Lösung [EMELJANOV (1976)] und der Aufspaltungsmethode [KELLOGG, TSAN (1978)]. Wesentlich sind die Untersuchungen zu Verfahren höherer Ordnung. Hier findet man auch einige Vermutungen zur Konstruktion von Verfahren beliebiger Ordnung. Abschließend wird erstmalig für Systeme ein gleichmäßig konvergentes Verfahren angegeben und das NEWTON-Verfahren für nichtlineare Probleme diskutiert.

Im 3. Abschnitt werden zunächst an drei Beispielen die Vorzüge angepaßter Verfahren gegenüber Standardverfahren demonstriert. Anschließend wird an einer Vielzahl von Problemen experimentell die Konvergenzordnung bestimmt. Dies ist interessant, weil die theoretisch abgesicherten Ordnungen oft nicht optimal sind. Abschließend werden angepaßte Verfahren an Wendepunktsproblemen getestet und ein FORTRAN-Programm für ein spezielles nichtlineares Problem bereitgestellt.

Insgesamt enthält das Buch eine Fülle von Ergebnissen und Anregungen zur weiteren Arbeit.

Magdeburg H. GOERING

Archetti, F. / Cugiani, M. (eds.), Numerical Techniques for Stochastic Systems. Conference, Gargnano Italy, 1979. Amsterdam-New York, North-Holland Publ. Co. 1980. VIII, 406 S., US $ 53.75. Dfl. 110.00. ISBN 0-444-86000-2

This volume contains 23 interesting and well-written papers, in which either stochastic problems are solved numerically or numerical problems are studied using stochastic ideas. The spectrum of the papers is very broad, and therefore this volume is of potential interest for many readers. Frequently occuring terms are: least squares, global optimization, stochastic optimization and earthquakes.

List of the papers: L. C. W. DIXON and P. J. LOVESEY: Nonlinear stochastic system identification viewed as an optimization problem; A. E. PEARSON: Equation error identification with modal disturbance suppressed; L. LJUNG: Fast algorithms for recursive estimation and identification; T. SÖDERSTRÖM: Spectral decomposition with application to identification; D. W. CLARKE: Some implementation: considerations of self-tuning controllers; N. BELLOMO: On a class of stochastic dynamical systems: mathematical analysis, some optimization problems and applications; P. D'ALLESANDRO: Classical theory vs. white noise theory; G. B. DI MASI and W. J. RUNGGALDIER: A continuous-time approximation to optimal non-linear filtering; C. BONIVENTO et al.: The use of stochastic models in air pollution prediction and control; E. CANUTO et al.: Lower order model representation of multivariable systems by factorization techniques; E. G. GUAGENTI: Stochastic processes in the modelling of earthquake occurrence; B. BACCHELLI and M. DI NATALE: General purpose versus specialized algorithms for an ill-conditioned estimation problem; E. SPEDICATO: Parameter estimation and least squares; C. CORRADI and L. STEFANINI: Efficient computation of maximum likelihood estimates in ARMA regression models; M. CUGIANI: A stochastic analysis of errors in numerical processes; F. ARCHETTI: Analysis of stochastic strategies for the numerical solution of the global optimization problem; J. J. MCKEOWN: Aspects of parallel computation in numerical optimization; B. BETRÒ: Experiment design for nonlinear regression; S. D'ATRI: Outline of a probabilistic framework for combinatorial optimization; J. MOCKUS: The simple Bayesian algorithm for the multidimensional global optimization; R. ZIELINSKI: Global stochastic approximation; C. G. E. BOENDER et al.: Global optimization: a stochastic approach; J. KORONACKI: Some remarks on stochastic approximation methods.

Freiberg D. STOYAN

Mercier, B., Lectures on Topics in Finite Element Solution of Elliptic Problems, Notes by Vijayasundaram, G., Berlin-Heidelberg-New York, Springer-Verlag and Bombay, Tata Institute of Fundamental Research 1979, V, 191 S., 30 Abb., DM 16,—. US $ 9.50. ISBN 3-540-09543-8

In dieser Schrift liegt die Ausarbeitung einer Vorlesungsreihe über „Finite Elementapproximation" vor. Nach Einführung in die Grundlagen der SOBOLEV-Räume und der abstrakten Variationsmethoden wird die Methode der finiten Elemente dargestellt mit Beispielen aus Elastizitätstheorie und Hydrodynamik. Ferner wird auf die Verfahren zur rechnerischen Lösung der entsprechenden approximativen Probleme eingegangen. Danach wird ein Überblick gegeben über die Fehlerabschätzungen bei diesen Methoden. Probleme mit einschränkenden Bedingungen für den Definitionsbereich werden betrachtet, danach gemischte Methoden: Fortsetzungsproblem, Fehlerabschätzungen mittels dualer Methoden, Aposteriori-Fehlerabschätzungen, Anwendungen auf die Gleichung von NAVIER-STOKES, die biharmonische Gleichung und das Plattenproblem, Methoden von HERMAN-JOHNSON und MORLEY. Es wird weiter die näherungsweise Lösung von Eigenwertproblemen mittels der Finiten Elementapproximation beschrieben.

Schließlich wird auf nichtlineare Probleme eingegangen, die in Zusammenhang stehen mit Variationsproblemen, es werden Existenzfragen und approximative Lösungen begründet. Der Leser erhält durch das Studium dieser Schrift einen guten Einblick in die mannigfaltigen Aspekte der Finiten Elementapproximation.

Berlin E. HEYN

Schüßler, H. W., Netzwerke, Signale und Systeme. Band 1: Systemtheorie linearer elektrischer Netzwerke, Berlin-Heidelberg-New York, Springer-Verlag 1981. XI 382 S., 210 Abb., DM 54,—. US $ 28.40. ISBN 3-540-10524-7

Das auf zwei Bände veranschlagte Lehrbuch will Studenten der Elektrotechnik in die Grundlagen der Systemtheorie elektrischer Netzwerke einführen.

Im vorliegenden ersten Band wird die Analyse linearer Netzwerke behandelt. Dabei wird besonderer Wert auf eine weitgehend systemtheoretische Betrachtungsweise gelegt, was unter anderem in der frühzeitigen Einführung der Übertragungsfunktion einer komplexen Frequenzvariablen und der ausführlichen Diskussion ihrer Eigenschaften zum Ausdruck kommt.

Zunächst wird die Analyse von linearen Gleichstromnetzwerken betrachtet. Gegenstand sind dabei insbesondere die KIRCHHOFFschen Gesetze sowie die Maschen- und die Knotenanalyse. In bewährter Weise werden dann die hier gewonnenen Erkenntnisse auf die Analyse allgemeiner linearer Wechselstromnetzwerke erweitert, wobei zunächst mit exponentiellen Zeitfunktionen gearbeitet wird. Aus den dabei angestellten Überlegungen werden einige allgemeine Sätze der Netzwerktheorie hergeleitet. Nach einer kurzen Darstellung der Vierpoltheorie werden die Eigenschaften der Übertragungsfunktion anschaulich erläutert. Es wird dabei unter anderem auf Mindestphasensysteme und Allpässe, Bode-Diagramme, Ortskurven und auf algebraische Stabilitätstests eingegangen. Das letzte Kapitel gibt eine umfassende Beschreibung der Einschwingvorgänge und betrachtet dabei auch die Aufstellung und Lösung der Zustandsgleichungen von Netzwerken sowie die Behandlung der Einschwingvorgänge mittels der LAPLACE Transformation' wobei durch letztere insbesondere eine Verbindung zur Wechselstromrechnung hergestellt wird. Der Anhang enthält eine kurze Darstellung der physikalischen Eigenschaften passiver und aktiver realer Bauelemente einschließlich ihrer Ersatzschaltungen sowie eine knappe, verständliche Erklärung der verwendeten mathematischen Grundlagen (u. a. der LAPLACE Transformation).

In diesem den dargebotenen Stoff induktiv aufbauenden Lehrbuch werden viele Überlegungen mit der Betrachtung oftmals auch praktisch interessanter Beispiele begonnen, wobei über die Spezifik hinaus allgemeine Zusammenhänge gesucht werden. Eine große Anzahl der Beispiele werden noch durch Meßkurven veranschaulicht. Auf eine kurze Literaturliste am Ende eines jeden Kapitels wird gezielt verwiesen

Dieses sich durch eine einheitliche Darstellung auszeichnende, gut lesbare Lehrbuch kann nicht nur Studenten der entsprechenden Semester des Elektrotechnik-Studiums, sondern auch anderen, an einem Zugang zu diesen Problemen interessierten Studenten sehr empfohlen werden.

Berlin J. BORCHARDT

Nussbaumer, H. J., Fast Fourier Transform and Convolution Algorithms. Springer Series in Information Sciences 2. Berlin-Heidelberg-New York, Springer-Verlag 1981. X, 248 S., 34 Abb., DM 69,–. US $ 40.80. ISBN 3-540-10159-4

Die Entwicklung der schnellen Algorithmen, die für die moderne Rechentechnik typisch sind, begann 1965 mit der grundlegenden Arbeit von COOLEY und TUKEY über die schnelle FOURIER-Transformation und wurde durch die Arbeiten des Verfassers des vorliegenden Buches wesentlich vorangetrieben. In Auswertung von fast 100 Originalarbeiten, die vor allem in den IEEE Trans. ASSP erschienen sind, bietet das Buch erstmals eine einheitliche Darstellung dieser Algorithmen auf der Grundlage von Polynomalgebren, wobei die zum Verständnis erforderlichen zahlentheoretischen Grundlagen in einem besonderen Kapitel bereitgestellt werden. Der enge Zusammenhang mit den Faltungen wird laufend ausgenutzt, und die verschiedenen Polynom-Transformationen und zahlentheoretischen Transformationen werden in ihrer Wirkungsweise verglichen. Wegen der erreichten theoretischen Durchdringung, der großen Bedeutung für die Rechentechnik und der weitergehenden Anwendungsmöglichkeiten in der Physik und der Elektrotechnik ist dem gut geschriebenen Buch ein breiter Leserkreis unter den Fachleuten und Studenten zu wünschen.

Rostock L. BERG

Mayr, R., Automatisierte Netzgenerierung für Finite-Element-Verfahren. Produktionstechnik Berlin 20. München-Wien, Carl Hanser Verlag 1981. 270 S., ISBN 3-446-13308-9

Für die Methoden der Finiten Elemente ist auf Grund hoher Anpassungsfähigkeit an unterschiedlichste geometrische Strukturen ein weites Anwendungsfeld erschlossen worden. Die Nutzung dieser Variabilität ist eine zunehmende Herausforderung an die Informationsverarbeitung, denn die Beschreibung der Geometrie komplizierter Strukturen und ihre sachgerechte Zerlegung in finite Elemente bringen in der Regel einen Aufwand mit sich, der den der eigentlichen Rechnung weit übersteigt. Die teilweise Automatisierung der Netzgenerierung für finite Elemente im Zusammenhang mit anderen durch CAD-Systeme bereitgestellten Mitteln ist sicher eine der Aufgaben, die für einen verstärkten Einsatz der Methode der finiten Elemente z. B. für Konstruktionsberechnungen im Maschinenbau zu lösen sind.

Der Autor des vorliegenden Buches gibt anhand eines speziellen Systems einen abgerundeten Überblick über die in diesem Zusammenhang zu lösenden Probleme.

Diskutiert werden die Generierung von Flächen- und Raumnetzen und die Glättung irregulärer Netze mit Hilfe von Optimierungsverfahren. Wesentlich erscheint, daß die rechnerinterne Darstellung von Netzen, die graphische Darstellung und die Manipulierung von FEM-Netzen als speziellen geometrischen Objekten mit gleicher Sorgfalt dargestellt werden, so daß das Buch von allen mit der praktischen Anwendung der Methode der finiten Elemente Befaßten mit Interesse aufgenommen werden wird.

Leider hat der Autor einen in einer solchen Spezialarbeit nicht unbedingt erforderlichen Abschnitt aufgenommen, in dem er die Methode der finiten Elemente erklären will. Dabei wird insbesondere eine krasse Mißachtung der gesamten Literatur zur Konvergenz der finiten Elemente deutlich. So ist sicher auch zu erklären, warum die für das Anliegen des Buches immanent wichtigen Erkenntnisse über Fehlerschätzungen und ihre Nutzung zur Anpassung von FEM-Netzen an die zu approximierende Lösung keine Aufnahme gefunden haben.

Berlin G. LIPPOLD

Peterson, J. L., Computer Programs for Spelling Correction: An Experiment in Program Design. Lecture Notes in Computer Science 96. Berlin-Heidelberg-New York, Springer-Verlag 1980. VI, 213 S., 8 Abb., DM 25,–. US $ 14.80. ISBN 3-540-10259-0

Programme zur automatischen Korrektur der wohl unvermeidlichen „Tippfehler" gehören bereits zu den klassischen Projekten im nichtnumerischen Bereich. Die vom Autor zusammengestellte Bibliographie weist Lösungsansätze aus mehr als zwei Jahrzehnten nach. Sie zeigen eine erstaunliche Vielfalt in bezug auf das für die Fehlererkennung genutzte systeminterne Referenzwissen und die mit unterschiedlichen Qualitätserwartungen implementierten Korrekturalgorithmen. Diese Ansätze werden im ersten Teil des Buches einer kritischen Analyse unterzogen und nach Leistungsfähigkeit und Aufwand bewertet. Das Schwergewicht liegt dabei auf Systemen für die interaktive Fehlerkorrektur, wodurch Fragen der Rechenzeiteffektivität immer wieder eine zentrale Rolle spielen. Die diskutierten Möglichkeiten zu ihrer Erhöhung reichen dabei von der Berücksichtigung sprachstatistischer Gesetzmäßigkeiten bis hin zur Ausnutzung von Wortbildungsmechanismen in einer Präfix- und Suffixanalyse, um auf diese Weise die Wörterbuchgröße minimal zu halten.

Entsprechend der vom Autor vorgegebenen Zielstellung besteht die Hauptaufgabe des Werkes allerdings nicht in der Beschreibung eines neuen Fehlerkorrektursystems, sondern vielmehr in der Illustration einer zeitgemäßen Implementierungstechnologie am Beispiel einer nichttrivialen Problemstellung. Deshalb enthält der weitaus umfangreichere zweite Teil die minutiöse Darlegung des gesamten Entwurfprozesses von der Systemaufgabe bis zum vollständigen Quellprogramm auf der Basis einer PASCAL-ähnlichen Notation. Die Darstellung ist ausgezeichnet aufbereitet, wobei die Wahl selbstkommentierender Variablen- und Prozedurnamen sowie ausführliche Kommentare und Begründungen zu den einzelnen Teilschritten das Verständnis sehr erleichtern. Im Gegensatz zu verschiedenen anderen Software-Dokumentationen fügt sich hier der hohe Anteil an Programmkode organisch in den Gesamttext ein und beeinträchtigt die durchgängige Lesbarkeit und das Vergnügen an der Lektüre in keiner Weise.

Die Erarbeitung der Programme erfolgt streng nach dem „top-down"-Prinzip der schrittweisen Verfeinerung. Dabei werden neben den entscheidenden Vorzügen dieser leistungsfähigen Methode an einigen Punkten aber auch ihre Schwächen sichtbar: Die Motivation für bestimmte Komponenten auf einem höheren Programmniveau wird teilweise erst durch ihre Implementation auf der niedrigeren Ebene vollständig klar. In diesem Sinne ist der hier dargelegte Entwurfsprozeß sicherlich idealisiert, während in der Realität, speziell unter den Bedingungen eines experimentellen Programmentwurfs, noch zusätzliche Rückkopplungszyklen mit den entsprechenden Veränderungen am Programmkode erforderlich sein dürften.

Originell wurde vom Autor die Frage der Arbeitsorganisation bei der Implementierung gelöst. So entstand der kompilierungsfähige Quelltext durch automatisches Streichen deru erläutern den Zwischentexte (und eine anschließende mannelle Nachredaktion) aus dem bereits vorliegenden, maschienlesbaren-Buchmanuskript!

Insgesamt hinterläßt das Buch beim Leser ein gewisses Erstaunen darüber, welche außerordentliche Durchsichtigkeit und Verständlichkeit für Software — auch in den Details — durch eine logisch klare Darstellung ihrer top-down-Entwicklung erreicht werden kann. Für den vom Autor vorgeschlagenen Verwendungszweck als Lehrbeispiel zur Anwendung moderner Programmierprinzipien scheint es daher auch bestens geeignet zu sein.

Berlin W. MENZEL

Noltemeier, H. (ed.), Graphtheoretic Concepts in Computer Science. Proceedings, Bad Honnef, 1980. Lecture Notes in Computer Science 100. Berlin-Heidelberg-New York, Springer-Verlag 1981. X, 403 S., DM 41,50. US $ 21.80. ISBN 3-540-10291-4

This volume contains the proceedings of the sixth Workshop on 'Graphtheoretic Concepts in Computer Science'. These concepts seem to play an increasingly important role in computer science as well as in a huge variety of applications.

The proceedings include 30 contributions on graphtheoretic models, complexity, net theory, combinatorial and optimization problems.

Special interest is centered on data base design and data structures in the framework of graph theory (4 contributions each). Although most of the contributions are subject to a more theoretical treatment, there are also some articles devoted to implementational aspects of algorithms working on graphs.

The value of these proceedings consists in the fact that results presented by leading scientists give an idea of the current state of the art in the field of Graphtheoretic Concepts in Computer Science.

Berlin M. HARTWIG

Ben-Israel, A. / Ben-Tal, A. / Zlobec, S., Optimality in Nonlinear Programming. A Feasible Directions Approach. New York-Chichester-Brisbane-Toronto, John Wiley & Sons 1981. XII, 144, S., £ 12.50. ISBN 0-471-08057-8

Ausgehend von dem Konzept zulässiger und brauchbarer Richtungen werden im vorliegenden Buch unterschiedliche Optimalitätskriterien für endlichdimensionale Optimierungsprobleme zusammengestellt und in ihrer wechselseitigen Abhängigkeit untersucht. Ein besonderer Schwerpunkt der Ausführungen liegt, durch die originalen Beiträge der Autoren hierzu stimuliert, in der Abschwächung der Regularitätsforderungen an das Ausgangsproblem durch Einbeziehung der Konstantheitsrichtungen (directions of constancy). Insgesamt gliedert sich das Buch neben der Unterteilung in konvexe und nichtkonvexe Probleme in die folgenden 5 Abschnitte: Charakterisierung von Optimallösungen; einige numerische Verfahren; ausgewählte Anwendungen; allgemeine notwendige Bedingungen; Optimalitätsbedingungen für differenzierbare Programme. Neben konvexen Problemen wird auch kurz auf einige Verallgemeinerungen des Konvexitätsbegriffes eingegangen und das dabei zugrundeliegende Prinzip anhand einfacher Beispiele verdeutlicht. Zur Ableitung der Beziehung zwischen Optimalitätsbedingungen vom Typ der zulässigen Richtungen und Bedingungen vom KUHN-TUCKER-Typ werden Trennungs- und Alternativsätze bereitgestellt und angewandt.

Ausgehend von Charakterisierungen der Optimallösungen unter Verwendung der Konstantheitsrichtungen werden auch für die Verfahren der zulässigen Richtungen adäquate Modifikationen untersucht. Für spezielle Aufgabenklassen gelingt es dabei, realisierbare Techniken zur Bestimmung der Konstantheitsrichtungen anzugeben. Implementierbare Schrittweitenalgorithmen ergänzen die Ausführungen zu Optimierungsverfahren. Charakterisierungen von PARETO-Optima sowie von Lösungen diskreter Minimaxprobleme werden als spezielle Anwendungen der vorher dargelegten allgemeinen Kriterien gegeben. Bei der Untersuchung nichtkonvexer Programme konzentrieren sich die Autoren vor allem auf die Einbeziehung von Kriterien 2. Ordnung.

Insgesamt kann eingeschätzt werden, daß mit dem vorliegenden Buch eine nützliche Zusammenfassung neuerer, durch die Konstantheitsrichtungen beeinflußter Untersuchungen zur Optimierungstheorie erfolgte, die durch ihre übersichtliche und klare Darstellung sicher gern von Mathematikern und Anwendern der Optimierungsalgorithmen zur Hand genommen werden wird.

Dresden C. GROSSMANN

In der DDR beziehbar im Rahmen des Kontingents.

EINGEGANGENE BÜCHER

Bei der Schriftleitung sind folgende Bücher eingegangen (erforderliche Besprechung bleibt vorbehalten).

Rall, L. B., Automatic Differentiation: Techniques and Applications. Lecture Notes in Computer Science 120. Berlin-Heidelberg-New York, Springer-Verlag 1981. VIII, 165 S., DM 21,50. US $ 9.80. ISBN 3-540-10861-0

Herman, G. T., Image Reconstruction from Projections. The Fundamentals of Computerized Tomography. New York-London, Academic Press 1980. XII, 316 S., US $ 29.50. ISBN 0-12-342050-4

Glaeser, G., Mathematik für Lehrer in Ausbildung und Praxis. Braunschweig-Wiesbaden, Vieweg 1981. X, 165 S., 34 Abb., DM 29,80. ISBN 3-528-08420-0

Ruzsa, I., Modal Logic with Descriptions. Budapest, Akadémiai Kiadó. 1981. 135 S., Ft 120,–. ISBN 9630522330

Brebbia, C. A. (ed.), Boundary Element Methods. Proceedings of the 3d International Seminar, Irvine, California 1981. Berlin-Heidelberg-New York, Springer-Verlag 1981. XXIV, 622 S., 232 Abb., DM 128,–. US $ 58.20. ISBN 3-540-10816-5

Gräff, E. / Klempt, E. / Werth, G. (eds.), Present Status and Aims of Quantum Electrodynamics. Proceedings, Mainz 1980. Berlin-Heidelberg-New York, Springer-Verlag 1981. VI, 302 S., DM 38,–. US $ 17.80. ISBN 3-540-10847-5

West, B. J., On the Simpler Aspect of Nonlinear Fluctuating — Deep Water Gravity Waves (Weak Interaction Theory), Lecture Notes in Physics 146. Berlin-Heidelberg-New York, Springer-Verlag 1981. VI, 341 S., DM 33,–. US $ 15.00. ISBN 3-540-10852-1

Böhme, G., Algebra. Vierte, verbesserte Auflage. Anwendungsorientierte Mathematik 1. Berlin-Heidelberg-New York, Springer-Verlag 1981, XI, 406 S., 211 Abb., DM 36,–. US $ 16.40. ISBN 3-540-10492-5

Niemann, H., Pattern Analysis. Informations Sciences 4. Berlin-Heidelberg-New York, Springer-Verlag 1981. XIII, 302 S., 112 Abb., DM 74,–. US $ 33.70. ISBN 3-540-10892-4

Siekmann, J. H. (ed.), GWAI-81. German Workshop on Artificial Intelligence, Bad Honnef 1981. Informatik-Fachberichte 47. Berlin-Heidelberg-New York, Springer-Verlag 1981. XII, 317 S., DM 37,50. US $ 17.10. ISBN 3-540-10859-9

Zypkin, Ja. S., Grundlagen der Theorie automatischer Systeme. Berlin, VEB Verlag Technik 1981. 484 S., 362 Abb., M 65,–. BN 5529637. (Übersetzung aus dem Russischen)

Szász, F. A., Radicals of Rings. Disquisitiones Mathematicae Hungaricae 11. Budapest, Akadémiai Kiadó 1981. 287 S., Ft 250,–. ISBN 963052226 X

NACHRICHTEN

Josef Heinhold zu seinem 70. Geburtstag

Am 4. 7. 1982 jährt sich zum 70. Male der Geburtstag von Herrn Prof. Dr. JOSEF HEINHOLD, ein Anlaß zu einem Rückblick auf das vergangene Leben, seine Erfolge und Höhepunkte, und auch zu einem Ausblick auf Pläne für weiteres wissenschaftliches Wirken. In dieser Zeitschrift, bei welcher Herr Kollege HEINHOLD seit vielen Jahren als sehr aktives und geschätztes Mitglied des Herausgeberstabes gewirkt hat, geben wir einen kurzen Rückblick.

JOSEF HEINHOLD wurde am 4. 7. 1912 in Düsseldorf geboren, kam aber schon frühzeitig nach München, wo er 1918–1931 die Schule besuchte, Mathematik und Physik an der Universität studierte, um in diesen Fächern 1935 nach einem vorbildlich kurzen Studium von nur 4 Jahren das Staatsexamen für das Lehramt an höheren Schulen abzulegen. Ein Jahr war er dann Referent an der Luitpold Oberrealschule in München, wo er das 2. Staatsexamen ablegte. Er war dann Assessor an der Oberrealschule Ingolstadt, Assistent an der TH München bei BALDUS und FABER, promovierte 1938 an der Universität München bei PERRON und SOMMERFELD und habilitierte sich 1940 an der TH München für reine und angewandte Mathematik bei FABER und BALDUS. Von dem von ihm sehr verehrten Lehrer OSKAR PERRON, bei dem er mit Arbeiten über „Verallgemeinerungen und Verschärfung eines Minkowskischen Satzes" promovierte, erhielt er Anregungen zur „Geometrie der Zahlen", einem Gebiet, mit dem sich auch seine Habilitationsarbeit beschäftigt. Herr FABER förderte sein Interesse für praktische Verfahren der konformen Abbildung und Herr BALDUS für Integriermaschinen und Rechenanlagen. So war Herr HEINHOLD bei seiner großen Vielseitigkeit die geeignete Persönlichkeit zur Besetzung eines 1950 neu errichteten Lehrstuhls für angewandte Mathematik und mathematische Statistik an der TH München. Zur Statistik und zum Operations Research kam Herr HEINHOLD durch praktische Problemstellungen der Maschinen- und Bauingenieure und der Betriebswirte.

Das genannte 1950 neu gegründete Institut entwickelte sich unter der Leitung von Herrn HEINHOLD rasch zu einem in der ganzen Welt bekannten Lehrstuhl mit einer Reihe verschiedener Schwerpunkte aus dem Gesamtgebiet der angewandten Mathematik, von denen auch viele Einzelveröffentlichungen aus der jetzt 83 Nummern zählenden Publikationsliste von Herrn HEINHOLD und eine größere Anzahl von Lehrbüchern zeugen. Eines seiner Lieblingsgebiete war und ist die konforme Abbildung. Bei seinem ausgesprochenen Sinn für die Anwendbarkeit interessierte er sich besonders für die praktische numerische Durchführung der konformen Abbildung, auch unter Verwendung elektronischer Analog-Rechenanlagen; das von ihm aufgestellte Schmiegungsverfahren wird auch nach seinem Namen benannt.

Große Verdienste erwarb sich HEINHOLD mit seinen Arbeiten über Hybridrechner und Hybridanlagen und deren Entwicklung, wobei es sich als wichtig erwies, neue mathematische Methoden zu entwickeln und die Hybridanlagen nach der technischen und der Software-Seite weiter auszubauen. Die internationale Anerkennung dieser Arbeitsrichtung am Institut fand ihren Ausdruck in der Wahl von Herrn HEINHOLD zum Vizepräsidenten der IMACS (International Association for Mathematics and Computing in Simulation).

Von großer Bedeutung erwies sich auch der Ausbau der mathematischen Statistik. Die moderne Massenfabrikation verlangte ständig wachsende Qualitätskontrolle und -lenkung, so daß auch eine geänderte Bezeichnung „Lehrstuhl für Angewandte Mathematik und Mathematische Statistik" gerechtfertigt erschien. Hier hat sich Herr HEINHOLD besondere Verdienste erworben, da nach dem zweiten Weltkrieg ein großer Nachholebedarf gegenüber den angelsächsischen Ländern bestand. Hand in Hand mit dem Ausbau der Statistik wurde auch das Gesamtgebiet des Operations Research stark gefördert.

Aus der über vierzigjährigen Vorlesungstätigkeit von Herrn HEINHOLD, die ihm stets Freude machte und nie eine Last war, gingen zahlreiche Lehrbücher hervor, die er teilweise mit Mitarbeitern zusammen abfaßte. Es seien besonders die Lehrbücher der Funktionentheorie, die eine 9. Auflage erlebte, die in 4. Auflage erschienene, zusammen mit Prof. GAEDE geschriebene sehr schöne „Ingenieurstatistik" und das mit Herrn KULISCH herausgegebene Buch über Analogrechnen genannt; die Ingenieurstatistik gilt an der TU München seit vielen Jahren bei den Studenten als „das Buch" über Statistik.

Alle diese Bücher pflegen in exakter und doch für den Ingenieur verständlicher Weise den Kontakt zwischen mathematischer Theorie und den Anwendungen, einen Kontakt, der in der heutigen Zeit einer ständig fortschreitenden Isolierung der Mathematik und einer Vernachlässigung des Konkreten gegenüber dem Abstrakten wichtiger denn je ist. Zusammen mit Prof. F. L. BAUER hat er das „Mathematische Praktikum" an der TU München aufgebaut und es zur Pflicht für die Mathematikstudenten gemacht, wodurch der Angewandten Mathematik ein großes Gewicht in der Mathematiker-Ausbildung beigemessen wird.

Zu seiner intensiven Lehrtätigkeit gehörte auch die Betreuung einer großen Zahl von Diplomanden, Doktoranden und Habilitanden; vor nicht langer Zeit erlebte Herr HEINHOLD seine 28. Dissertation als Doktorvater, und von seinen zahlreichen Habilitanden sind inzwischen viele selbst Lehrstuhlinhaber geworden, und andere Schüler von ihm sind inzwischen an leitender Stelle in Industrie- und Versicherungsunternehmen tätig.

Es konnte nicht ausbleiben, daß er auch Rufe an andere Hochschulen, so an die Technischen Hochschulen Hannover und Darmstadt erhielt; aber er blieb der Stadt München und seiner Hochschule treu. München war für ihn als Bergfreund, Skiläufer und Bergwanderer die Wahlheimat; dort pflegte er seinen Freundeskreis und auch seine Hobbys, von denen nur das Tischlern sehr schöner Möbelstücke, die er in einem gut ausgestatteten Hobbyraum anfertigt, erwähnt sei, und zu seinen Erholungen zählt auch das Segeln im eigenen Boot auf den bayerischen Seen. In dem Hause, welches seine Frau Renate ihm zu einem ruhigen Heim gestaltet, findet er die nötige Entspannung und Kraft zu neuer Arbeit. Sehr groß ist die Zahl seiner Freunde, die er sich im In- und Auslande durch sein stets freundliches, liebenswürdiges, hilfsbereites und humorvolles Wesen erworben hat. Mögen ihm noch viele Jahre erfolgreichen Lebens und Wirkens beschieden sein.

Hamburg L. COLLATZ

FSC
www.fsc.org
MIX
Papier | Fördert gute Waldnutzung
FSC® C083411

Zeitfracht Medien GmbH
Ferdinand-Jühlke-Straße 7
99095 Erfurt, Deutschland
produktsicherheit@kolibri360.de